GCHQ PUZZLE BOOK II

By the same author

The GCHQ Puzzle Book

GCHQ Puzzle Book II

PENGUIN BOOKS

PENGUIN BOOKS

UK | USA | Canada | Ireland | Australia
India | New Zealand | South Africa
Chameleon | Gibbon | Jackal | Kangaroo | Lizard
Penguin | Rhino | Scorpion | Tarantula | Vulture

Penguin Books is part of the Penguin Random House group of companies
whose addresses can be found at global.penguinrandomhouse.com.

First published 2018
004

Text copyright © Crown Copyright, 2018

The moral right of the author has been asserted
bellow dawning feast field helped learning
model mouthful ponder puzzle

Set in Plantin
Designed by Couper Street Type Co.
Printed and bound in Great Britain by Clays Ltd, Elcograf S.p.A.

A CIP catalogue record for this book is available from the British Library
delphi macrobiota mariachi multiply parvenu pepsi scampi taxi taffeta

ISBN: 978–0–241–36543–4
1 2 3 4 5 8 55 144 233 377
2 3 4 6 9 15 55 90 200 1010
3 6 7 10 21 22 30 33 35 36
1 4 5 8 10 23 46 61 84 5000

www.greenpenguin.co.uk

Contents

To the hard-working people
of GCHQ – past, present and future

I am immensely grateful to all those who bought the first GCHQ Puzzle Book. It raised hundreds of thousands of pounds for Heads Together, helping people of all ages get the vital mental health support they both need and deserve. I would like to thank the ingenious minds of GCHQ for creating the puzzles, even if I can't claim to have cracked them all!

It is an accepted part of life that we should keep physically fit, heading to the gym, out for a run, or playing a game of football. Sadly, we do not all put the same effort into keeping ourselves mentally fit. But just as physical activity is good for the body, regularly focussing on our mind can make an important difference in our lives. Whether it is taking some time out of our day to completely switch off, or to get lost in the puzzles of this book, keeping mentally fit as well as physically fit can play a valuable part of our routine.

Since it was formed in 1919, GCHQ's intelligence has protected members of the Armed Forces. When our troops are deployed overseas, I am reassured that some of the finest problem solvers in the world are working to keep them and other coalition forces safe. Although the men and women of GCHQ serve in secret, every day they play a pivotal role in the security of our country and the safety of our lives. Catherine is incredibly proud of the role her grandmother Valerie Glassborow played in the Second World War, as one of the code breakers at Bletchley Park. The role of GCHQ is as valuable now as it was then.

Over the last century, the world has changed dramatically and GCHQ's mission has had to adapt and evolve. What has not changed is the core of the organisation; people from all walks of life who work together to challenge convention and fulfil their potential.

All minds are vulnerable to the stresses and strains of everyday life and we welcome GCHQ's support in encouraging a more open conversation about mental wellbeing.

I hope you enjoy this book and thank you for supporting the work of Heads Together.

Foreword by Director GCHQ

When we launched our first Puzzle Book in 2016 we wanted to share the talents of the amazing people in GCHQ, inspire the next generation to test their puzzling skills, and raise funds for the mental health charity Heads Together. We knew it was good and hoped it would prove successful. However, none of us predicted the way in which it would capture attention, top the bestseller lists, and provide so much financial support for a brilliant cause.

It proved so successful that we decided to produce a second version. Our objectives remain the same. GCHQ has been at the heart of the nation's security for a hundred years. It is a world-class intelligence and cyber agency full of fantastically committed people using their ingenuity to keep the United Kingdom safe. We want to give you a further taste of how their minds work and how they go about solving the seemingly impossible. We have included some stories to bring this to life and, in so doing, we hope to provide greater insight into our history and the challenges we face as we approach our second century.

In GCHQ we know that even great minds need support. We are proud of the way in which GCHQ has always valued difference and different ways of thinking. We have a long history of nurturing talent. We understand that as the world changes and the pressures on us increase, we have a fundamental responsibility to create a supportive environment where we can be ourselves at work every day.

So, it is a particular privilege to support Heads Together, the mental health initiative run by the Royal Foundation of The Duke and Duchess of Cambridge, and The Duke and Duchess of Sussex, that spreads this message to organisations and individuals across the nation. Funding from the proceeds of this book will support this vital work.

I want to thank their Royal Highnesses for their patronage of this cause and their support to GCHQ. I would also like to thank all of those involved in creating this book. They have dedicated their time to this project on top of their important day jobs. For obvious reasons, they cannot be named, but I am in awe of all that they and the rest of the GCHQ team does to keep the country safe.

I wish you luck in tackling these puzzles. You will need to use all of your ingenuity to think around the problem. You may need to work in teams to get the answer. And you will have to draw on the reserves of your integrity to avoid peeking at the answers. In other words, you will have to deploy the same values we have in GCHQ to tackle the most serious cyber, terrorist, criminal and state threats. If at the same time we encourage you to briefly reflect on this essential work, and even to pause on whether you could be part of this mission, we will have succeeded again.

Thank you for taking up the challenge!

Jeremy Fleming
Director GCHQ
Cheltenham, August 2018

Since the First GCHQ Puzzle Book

Well, here we are again. Normally if a member of the intelligence agencies writes a bestselling book they are pursued by the courts – but in this case we were pursued by the publisher. Indeed after the publication of the first GCHQ Puzzle Book those of us most involved found ourselves thrown into the world of the media, not a place that GCHQ employees have traditionally found themselves!

So, first, thank you very much to all of you who bought *The GCHQ Puzzle Book*. By doing so, you propelled us into the bestseller lists and contributed to the £330,000 that we have donated to Heads Together in the first two years since publication. That money has helped continue the campaign to tackle stigma and change the conversation on mental health.

We now need to thank you again, of course, for buying *GCHQ Puzzle Book II*, and again contributing to the work of the charities that make up Heads Together.

There was a competition in the first Puzzle Book, and we were delighted that this got much attention. Most of the competition was in the book, but the answers led to a webpage address and a password, where there was a further puzzle. To get to this stage required a high level of puzzle-solving skill, and we were impressed that we had over 150 entries. The entrants who provided the best answer to the final puzzle then took part in a Tiebreaker, which was deliberately made extremely hard so that we could separate them! We have reproduced that Tiebreaker in this book (on pages 167–74) for everyone to enjoy!

The winner of the competition was Angus Walker. Second was Jessie Wheat. The competition prize was a wonderful glass trophy created by sculptor Colin Reid, which was deliberately based on the design of his Cipherstone – the art piece that can be found at the entrance to GCHQ's main building. Naturally we had Colin etch a puzzle into the glass – photos of the trophy, and close-ups of the puzzle, can be found in the second colour section. It was a delight to meet Angus and present him with this trophy last year.

It was the unprecedented level of interest in our Christmas Card Puzzle that led to the idea for *The GCHQ Puzzle Book*, and it was the fantastic success of that which has led to *GCHQ Puzzle Book II*. Like the first book, this one contains a whole range of different puzzles set by people who work at GCHQ.

We hope you enjoy it.

Introduction to the Puzzles

The puzzles in this book range in difficulty but we hope there is something for everyone. It can be hard to judge difficulty but we have attempted to do so, and throughout the book you will find Starter puzzles (marked by ♪), some Hard puzzles (marked by ♪) and everything else in-between is unmarked. Remember that in multi-part questions later parts tend to be harder than earlier parts. The Starter puzzles are intended to be easier than others, but that doesn't mean they are easy – so solving one is still an achievement!

For the Starter puzzles we have also provided clues for each question, which can be found on pages 199–203.

All the questions will become easier once you are tuned into the mindset of the setters – there are several repeated themes and styles that occur throughout. Because of this we have designed a way in which you can get help without cheating – by using the Hints section (on pages 175–82). In this we explain how to approach some questions and also give various themes and topics that you can consider if you are struggling with a puzzle. These are all presented as general hints, not specific to any particular question – so you can still give yourself full credit when you solve the puzzle you were stuck on.

For code puzzles you should read 'How to Solve a Cipher', which is on pages 183–97. This may also give you a taste of one of the aspects of GCHQ's work.

As an additional aid, particularly for those of you trying to answer questions whilst down a coal mine, in a jungle or on a desert island – in other words where you have no signal to access the internet – there is an appendix that contains some useful lists of information, and which, like the Hints section, may help you if you are struggling.

The puzzles themselves, as for the first Puzzle Book, are largely from our Christmas Puzzle Quiz and Kristmas Kwiz, but there are also many new questions written especially for this book. You will find a few Christmas references throughout, and also mention of the kwiz koffee kup. This is a notoriously unstable drinking vessel that tends to get knocked over and spill koffee over crucial parts of certain questions!

As this Puzzle Book has been produced to mark our centenary year, it contains a number of historical pieces describing aspects of GCHQ's history. After each of these you will find a puzzle with a connection to that subject. These puzzles are shown in boxes. We have also included examples of the actual entrance tests sat by applicants to GCHQ from the 1960s through to the 1990s. These examples include one I sat myself!

Also included is a set of puzzles we wrote to mark Her Majesty the Queen's ninetieth birthday (pages 161–66) and, if you relish a challenge, you can find the Tiebreaker we wrote to separate the first GCHQ Puzzle Book competition finalists on pages 167–74.

For a fresh puzzle challenge turn to the Centenary Trail, which starts in the second colour section.

We hope you enjoy the puzzles in this book, whether you start from the beginning and work through, or just dip in and out of the pages.

Puzzles

1. The early bird

What connects Skipper, Mumble, Wheezy, Pinga, Tux and Feathers McGraw?

2. Jane's Novelty Decorations

Last night Jane was decorating an unusual birthday cake. Her ingredients were 3oz marzipan, 6 juniper berries, 5oz mayonnaise – and 4 of what sort of fruit?

3. Bravo for the movies!

Complete:

(a) The Grand Budapest _____

(b) A Passage to _____

(c) The Treasure of the _____ Madre

(d) _____/Victoria

(e) _____ Doodle Dandy

(f) _____

a–e did. f didn't. What?

4. Divide into pairs I

Pair the following:

AN, BLACK, EX, LIVER, NEW, NOR, OX, PORTS, SOU, TOR, WAT, WEN

5. What the . . . ?

Which is the odd one out?

Bleak, Great, Hard, Little, Twist

6. A lot of effort!

Arrange into 7 groups of different sizes:

40km Road Cycling	100m Hurdles	200m
200m Freestyle	800m	1500m Freestyle
10,000m	Balance Beam	Cross-country Run
Cross-country Skiing	Elimination Race	Fencing
Floor	Flying Lap	High Jump
Individual Pursuit	Javelin	Long Jump
Pistol Shooting	Points Race	Rifle Shooting
Scratch Race	Shot Put	Show Jumping
Time Trial	Triple Jump	Uneven Bars
Vault		

7. A gimme

Does your mother know why:

Ring + Money = I Do ?

8. **Picture the numbers**

What image is formed by:

10		2
1	5	20
	50	

9. **Three's a crowd**

Arrange into pairs:

BLIND	CIRCUS	DAY	EVENT	FRENCH	HENS
ISLAND	LEGGED	LINE	MICE	MILE	MONKEYS
POINT	RACE	RING	TURN	WHIP	WISE

10. **Tickets**

If a train ticket from Wigan to Ripon costs £1, a ticket from St Ives to Tiverton costs £4 and a ticket from Hove to Sevenoaks costs £5; how much is a ticket from Poole to Aylesbury – and why might going by road seem more appropriate?

11. **Mssng vwls**

Divide the following items into 7 sets of 7.

BL	CLSSS	CNDL	FRC	FRDY	FRTN
FRTYNN	FRTYTW	GRN	GLTTNY	GRD	GRDNS
KLGRM	KLVN	LGHTHS	LST	ML	MNDY
MPR	MSLM	MTR	NDG	NRTHMRC	NTRCTC
NVY	PRD	PYRMD	RD	RNG	RP
S	SCND	SLTH	SNDY	STHMRC	STRDY
STRL	STT	SVN	THRSDY	THRTYFV	TMPL
TSDY	TWNTYGHT	TWNTYN	VLT	WDNSDY	WRTH
YLLW					

12. The pick of the litter

What connects:

A capital city in Oceania

A European country bordering Turkey

A river flowing through Colombia and Venezuela

A Channel Island whose capital is St Anne

13. Some sums

(a) If R+B=P and R+Y=O, then B+Y=?

(b) If R+B=M and B+G=C, then R+B+G=?

(c) If Y+Y=B, Y+G=B, R+P=B and G+B=B, then R+R=?

14. Odd one out I

What word or phrase is the odd one out in each group?

(a) Disappear, Grapple, Peace, Pondicherry, Sublime

(b) Crewe, Hootenanny, Kitchen, Nightmare, Tomato

(c) Firenze, Gilead, Jargon, Marigold, Pumpernickel

(d) Catwoman, Deus ex machina, Parishioner, Pyromania, Scuba

15. Pop and Pokémon

When we quizzed a group of musical artists about their favourite
Pokémon, there were no big surprises:

Daniel Merriweather	Charizard
Eiffel 65	Blastoise
Coldplay	Pikachu (they also liked Xerneas and Yveltal)
Spandau Ballet	Ho-Oh
Echo and the Bunnymen	Lugia
New Order	Suicune

but what answer did the Kaiser Chiefs give?

16. Superman

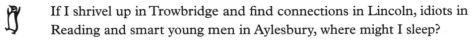

If I shrivel up in Trowbridge and find connections in Lincoln, idiots in Reading and smart young men in Aylesbury, where might I sleep?

17. Doors

Divide the following words into two sets:

AB, COP, FANCY, FLUX, HIDE, KING, RAG, VEST, WARD, WIPE

18. What's the answer?

(a) What is the first letter of the Greek alphabet?

(b) What sort of rain did Guns N' Roses sing about?

(c) What Spanish word means a range of mountains?

(d) What are Ardbeg, Glenfiddich and Talisker?

(e) Which nymph faded away until all that was left was her voice?

(f) Which Shakespeare character completes this answer?

19. Sequence I

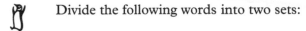

What word could follow:

SQUABBLE, ANECDOTE, WAVEFORM, TOUGHEST, DEMIJOHN, RECKLESS, ?

20. Chickens in the gallery

The dog's in the canal, the cat's on the fell, and the sheep is over there. Is the pig in the cellar, the larder or the porch?

21. Football logic puzzle I

In the following table every pair of teams has played each other once. The table is ordered in the usual way (3 for a win, 1 for a draw) with goal difference, goals for and fair play record used as tiebreakers. Using arithmetic and logic, fill in all the gaps.

Team	Points	Goals For	Goals Against	Goal Difference
Wanderers	5	3		
United			6	2
City	4			
Rovers	3		3	

United	1	v	3	Rovers
Wanderers		v	0	City
City		v		Rovers
Wanderers		v	2	United
City		v		United
Rovers		v		Wanderers

22. Lost animals

Replace the *s with animals to make words, e.g. exe*le → exeCRABle

(a) appr*imate

(b) boon*gle

(c) mead*ark

(d) bil*aire

(e) wilde*st

(f) che*erapy

(g) af*oon

(h) e*lottis

(i) lyc*hrope

(j) bene*nt

(k) st*agem

(l) interre*m

(m) b*zebub

(n) stac*o

(o) ver*im

(p) si*ure

(q) *orker

(r) mi*ave

(s) ex*ble

(t) cont*e

ORIGINS OF GCHQ AND ITS MISSION

GCHQ was founded in 1919 under the name of GC&CS, the Government Code and Cypher School, but its story has many beginnings.

The origins of UK Signals Intelligence (usually referred to as Sigint) can be found at the start of the First World War, although the principal techniques it uses – interception, cryptanalysis and traffic analysis – had been in use for centuries. The British government had no intelligence agencies before 1909, and the small organizations founded that year which would develop into MI5 and MI6 focused on people who could gather intelligence on German capabilities and intentions, and people who could counter the threat from spies for Germany who were trying to do the same thing in the UK. It would be another ten years before an agency dedicated to deriving intelligence from communications would be established.

Even though there was a rapid spread in wireless technology among the most advanced navies in the first decade of the 1900s, there was no appreciation that it might be possible to examine the communications of other navies to gather intelligence. There are various reasons for this: strategic thinking about wireless focused mainly on the

transformational capability of radio to allow, for the
first time, those in power ashore to manage ships beyond
sight of land; most serious technical attention was being
paid to developing and exploiting the new technology for
the Royal Navy's own use – so foreign transmissions were
a distraction; and, anyway, when Fleet Paymaster Charles
Rotter, the one officer in the Naval Intelligence Division
of the Admiralty who wondered about the potential of
interception, was allowed to use the Admiralty's own
wireless sites to carry out trial interception of German
naval signals, neither he nor anybody else was able to break
the codes the Germans were using to encrypt their messages.

The British Army's starting point was different. It
understood that the ability to decrypt a coded dispatch
captured from an enemy courier had the potential to
produce intelligence but thought of this as something of
immediate tactical value. The conclusion it drew was that
every officer should be taught the basics of cryptanalysis.
Some modest success during the Boer War confirmed this
view, though for various reasons it was only in India that
a two-man unit was set up to create working aids to allow
officers to decode Russian, Chinese and Tibetan messages.
The mobilization plan for the War Office in London assumed
that, in wartime, this function would be managed as a
secondary part-time responsibility for the War Office
Librarian and the Director of Army Sanitation.

During the lead-up to the outbreak of war, in July 1914 the
widespread use of wireless by the German Army and Navy to
command and control their mobilization and subsequently
the first offensive operations came as a shock to the
British. It was relatively straightforward to build up an

interception capability: that was simply an organizational question of finding or producing wireless sets (the Admiralty's existing equipment was needed to support the Royal Navy, and the Army barely had enough for its own limited use), siting them to collect German messages and training the manpower needed to operate them. The sticking point was that the UK had no cryptanalysts to make sense of the encrypted messages.

The problem was solved by the generous intervention of foreign partners. In the case of the Army it was the French. France had been carrying out Sigint against the German Army for some years and was happy to teach its allies everything it knew. By September 1914 there was a leap forward in breaking the main German Army system and the War Office developed a system in which the General Headquarters on each Front (the Western Front, Egypt and Mesopotamia) had its own interception and cryptanalysis organizations, loosely coordinated by a section in the War Office eventually known as MI1(b). This section also looked at enciphered foreign telegrams carried by British cable companies. Out of it grew MI1(e), which supplied intelligence to the air defence system that began to counter the threat from German airships in 1915.

When a German battleship, the *Magdeburg,* ran aground in the Baltic on 26 August 1914, it was boarded by the UK's Russian allies, who were surprised to find two copies of the German High Seas Fleet codebook on board. It took a couple of months for one of these codebooks to reach the UK, and at first it yielded little to the staff in Room 40 of the Old Admiralty Building. However, Fleet Paymaster Rotter worked out that the Germans were adding a daily changing key to

the values in the codebook. Once this German technique had
been discovered, it was broken subsequently every day.

The Naval Sigint effort in Room 40 and the Army Sigint work
in MI1(b) continued to grow. From simple beginnings looking
at the German Navy and Army, British Sigint moved into the
interception of foreign diplomatic communications, and
the discovery of the techniques the Germans were using to
avoid Allied sanctions. By the end of the war, hundreds of
people were working in British Sigint, and thousands more
were using their output to inform decision-making across
the whole of government. The success of these efforts led
to what would become a permanent Signals Intelligence
organization in 1919 that would eventually develop into the
GCHQ of today.

23. A letter

Help! We can't work out who sent us this letter!

Dear GCHQ,

On the occasion of your centenary, we your forefathers thought it would be fitting to write. In our day, cryptologic work was prey to all sorts of catastrophe: lip pestilence and other diseases hampered our work; men of low ilk insisted that we were in league with the devil. Now all is a lot easier and we hope that those who follow will esteem both our work and yours. The torch we passed to you now heats to near-stellar temperatures. Your colleagues seem to have all our skill in mathematics and language but to judge by your puzzle books an equally deep knowledge of matters such as 70s music acts (the Brothers Gibb, ABBA, Genesis, for example).

Our best wishes for your future success

Thomas, John, John, Edward, Charles and Charles

24. More than one answer

What's the correct (and appropriate) answer? What isn't?

18.5, 27.2, 73.4, 19.3, 2.1, 33.6, 14.1, 42.2, 109.5, 34.4

25. Loads of words

If baron = 5, noon = 10, icon = 26 and ten = 50, how much is load?

26. Pot of gold

What comes next in this sequence: R, E, Y, N, B, O, ?

27. Multiple multiplications

Which multiplication has the highest answer?

(1) (1, 9, 8) × (2, 11, 19)

(2) (4, 1, 7) × (100, 5, 101, 10)

(3) (15, 2, 4, 3) × (50, 13, 5, 1)

(4) (15, 3, 20) × (5, 12, 7, 11, 9, 21)

28. Hadrian's birthday party

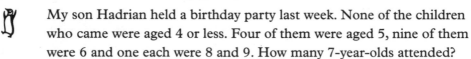

My son Hadrian held a birthday party last week. None of the children who came were aged 4 or less. Four of them were aged 5, nine of them were 6 and one each were 8 and 9. How many 7-year-olds attended?

29. Unfinished films

(a) A man is kidnapped and taken to the Balkan region for a little over a decade.

(b) A golfer protects a town against bandits.

(c) Chicago is terrorized by a neutered chicken.

(d) A detective investigates the case of a religious teacher falsely incriminated for murder.

(e) A scientist returns to an erupting volcano to retrieve a vegetable that once belonged to an Italian writer.

(f) After returning from combat in Iraq, a wading bird struggles to adjust to civilian life.

(g) While thwarting plans to build an underwater city, a veteran agent falls for his boss.

(h) At a protest during the hours of darkness, a psychologist investigates a satanic cult.

(i) A man attempts to obtain an eco-friendly vehicle by entering into a sham marriage.

(j)   Having acted in a litigious manner, a tall-haired man aims to become a rockabilly star.

(k)   A piano player turned lip-reading philanthropist competes at an ancient Chinese board game.

30.   Here comes the Sun

What are the missing letters?

SUN ?????SUN

31.   Next in line

Who is currently next?

C, W, G, C, L, H, A, B, E, E, J, L, A, P, S, I, Z, M, ?

32.   First and second

What comes first in the following sequences?

(a)   ? F M A M J J A S ...

(b)   ? E L N D J J R 1 ...

(c)   ? H L B B C N O F ...

(d)   ? B G D E Z E T I ...

(e)   ? D T Q C S S H N ...

What comes second in the following sequences?

(a)   A ? A P A U U U E ...

(b)   E ? E U E O U U S ...

(c)   Y ? I E O A I X L ...

(d)   L ? A E P E T H O ...

(e)   N ? R U I I E U E ...

33. An error

Which is in error?

DBMTQRK, HTRFJAA, JNFC, JNSQLOLIDSMK, LDPLHW,
NFSTQFET, NORTFKRN, OGUUBEGJJY, RBGNPHKK, VAGRMPAL

34. Divide into pairs II

Divide into pairs:

AGO, BIRD, BOY, CHIC, GIG, HAM, HOOD, I, LET, LIGHT,
LIVER, MAN, O, ON, PAT, PLATO, SPOT, TON

35. Mostly mixed up

Each of the following sets of letters has been created by taking 3
related words or names, removing the first letter from each, and
putting the rest of the letters in alphabetical order. Within each part
the letter removed was the same, and it doesn't appear among the
remaining letters. What were the 3 words or names in each case?

(a) eeehnorw

(b) aaelnnruuuyy

(c) aaaelnqrrrs

(d) adeinnooprxy

(e) aaabeehrssstt

In August 1914, the Royal Navy's embryonic Sigint operation in Room 40 of the Admiralty received valuable assistance in the shape of a copy of the main German naval code book, captured from the German cruiser *Magdeburg*. Many other German naval code books and cipher keys were captured or otherwise acquired throughout the war.

The first aerial bombing raids on Britain were carried out by German airships in 1915. British Sigint stations belonging to the Army and Navy could intercept coded radio transmissions made by the airships, tracking their location and enabling advance warning to be given to defence forces. Pictured is a decrypt of one such airship signal, handwritten by the duty cryptanalyst Alastair Denniston.

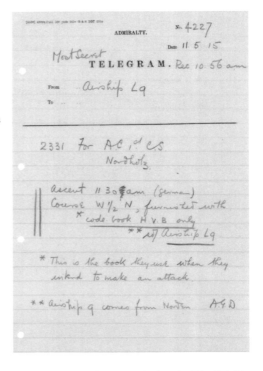

A leading cryptanalyst in Room 40 during the First World War, Denniston was the successful candidate to head GCHQ when it was formed on 1 November 1919 under Admiralty control, and known then as the Government Code and Cypher School (GC&CS). He would remain a key figure in British Sigint for the next two decades.

Text:

00424	00191	00437	00416	13045	17214	06251	06491
18147	19062	22295	17694	21141	04305	11580	39689	11313
10387	21604	23845	12137	17078	06015	16127	20438	22464
23612	16158	11836	09427	22927	09758	12137	22911	01245
15905	13247	07638	74038	06993	39689	11291	22464	05884
01240	22047	13372	06015	16127	52262	22394	17802	22284

Remarks

Text:

22464	13239	17257	01334	22309	22776	22931	22167
21995	18245	09427	12137	02985	17149	11664	04278
21668	23724	67893	04473	52262	14217	12137	14737
11313	09325	20362	15047	18812	97556

AFFAIRES ETRANGERES.

Remarks SWEDISH CYPHER.

A copy of the Zimmermann Telegram, forwarded after the event to the War Office censor by the intercepting cable company. Note the decrypted German plain text written above the original cipher groups. Opposite is another copy of the message, fully deciphered and translated by Nigel de Grey of Room 40. Germany never realised that its diplomatic codes were readable in this period.

"morning, J.O.!"

The Director of Naval Intelligence, Reginald 'Blinker' Hall, agreed with de Grey in immediately recognising the political dynamite the Zimmermann Telegram would represent in America. This caricature of Hall greeting his staff in the Intelligence Office is by G P Mackeson of Room 40.

FROM:- Washington
TO:- Mexico   No. 3

The Foreign Office telegraphs on Jan. 16.

We intend to begin on the first of February unrestricted submarine warfare. We shall endeavour in spite of this to keep the U.S.A. neutral. In the event of this not succeeding we make Mexico a proposal of alliance on the following basis:-

MAKE WAR TOGETHER

MAKE PEACE TOGETHER

Generous financial support and an understanding on our part that Mexico is to reconquer the lost territory in Texas, New Mexico and Arizona. The settlement in detail is left to you. You will inform the President of the above most secretly, as soon as the outbreak of war with the U.S.A. is certain and add the suggestion that he should on his own initiative invite Japan to immediate adherence and at the same time mediate between Japan and ourselves.

Please call the President's attention to the fact that the ruthless employment of our submarines now offersthe prospect of compelling England in a few months to make peace.

Acknowledge receipt.

ZIMMERMANN.

The All Russian Co-Operative Society (ARCOS) was the body responsible for the orchestration of Anglo-Soviet trade in the early days of Communist Russia. In 1927, its London offices were raided, the Baldwin government accusing the USSR of using the organisation for the conduct of espionage and subversion. Diplomatic and trade relations were suspended as a result. This Russian telegraphic code book is believed to have been captured during the raid and sent to GC&CS.

When he joined Bletchley Park in September 1939 the mathematician Alan Turing requested he be given the most difficult cryptanalytical problem to work on. He became head of Hut 8, responsible for breaking German naval Enigma, and developed the concept of the Bombe machine.

The factory floor of the British Tabulating Machine Company at Letchworth in Hertfordshire, where Bombe machines designed to break Enigma were manufactured. In the foreground are new wheels waiting to be fitted.

Hut 6 dealt with German army and air force Enigma. Pictured are the staff of the Hut 6 'Duddery', so named because their job was to examine messages which had not been successfully decrypted ('duds'), despite recovery of the Enigma keys involved in their encipherment. The Enigma machine in the foreground was used for checking traffic against solved cipher settings.

So named by their female 'Wren' operators because they were reminded of the fantastically overwrought designs of cartoonist William Heath Robinson, the series of 'Robinson' machines was designed to automate elements of the attack on the high-level German cipher Lorenz. This is an 'Old Robinson', on which two streams of paper tape are compared at high speed.

Building on insights gained from the Robinson machines and the design of electronic telephone exchanges, Post Office engineer Tommy Flowers led the team which developed Colossus, a complex valve-based machine and the world's first electronic computer. Ten machines were built, providing vital intelligence in the closing years of the war. A later example is seen here at Eastcote in 1949.

As a result of the decision by US President Roosevelt to expand cooperation with the UK, a team of American cryptanalysts arrived at Bletchley Park early in 1941 where they learned of British Sigint advances. Denniston followed this up later in the year with a dangerous return trip to Washington during which he established the foundations of the cryptographic partnership between the UK and the US which endures to this day. Pictured is his diary at the time.

Institutional partnerships between GCHQ and its allies have often been mirrored at a personal level. As a gesture of friendship in the face of shortages over the bleak winter of 1949, Christmas food parcels were collected and donated to junior GCHQ workers by staff at what became the US National Security Agency.

Scarborough on the North Yorkshire coast is the location of what is believed to be the world's oldest continually operating Sigint site. It began as a Royal Navy wireless telegraphy station as early as 1912 and became an intercept site two years later. This is a view of the site's bunkers, huts and aerials as they appeared in the 1960s.

GCHQ's satellite dish 'farm' at Morwenstow on the North Cornish coast near Bude.

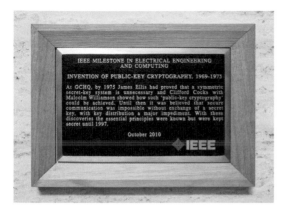

In 2010 the New York-based Institute of Electrical and Electronics Engineers (IEEE) recognized the pioneering secret work done by James Ellis, Clifford Cocks and Malcolm Williamson at GCHQ in the late 1960s and early 1970s which led to the invention of public-key cryptography – a principle which lies at the heart of all secure transactions on the internet.

GCHQ's iconic 'Doughnut' headquarters in Cheltenham. In 2014, to commemorate the centenary of the outbreak of the First World War, GCHQ teamed up with the Royal British Legion to launch its Poppy Appeal in Gloucestershire. Over 1400 staff, both civilian and military, worked together to create a giant poppy in the centre of the building.

Information security has long been an important and integral function of GCHQ's mission. In October 2016 the National Cyber Security Centre was set up as part of GCHQ. In February 2017 Her Majesty The Queen opened the NCSC's headquarters at Nova South in London.

The NCSC has helped to nurture the next generation of cyber security experts through its CyberFirst courses, enjoying great success inspiring young women with the CyberFirst Girls' competition. Pictured hard at work in Lancaster House are the finalists of the 2017 competition.

GCHQ's headquarters lit in rainbow colours in 2015 to show their commitment to diversity for IDAHOBiT day, which aims to raise awareness of issues around homophobia, transphobia and biphobia.

Could you hone your puzzling skills and be part of GCHQ's mission as it moves forward into its next hundred years?

www.gchq.gov.uk

36. Missed connections

What (specifically) connects:

37. One thousand

What's the connection?

(a) Home of Winnie-the-Pooh and friends

(b) Eight stone

(c) Share index of companies listed on the London Stock Exchange with the highest market capitalization

(d) Conflict between the House of Plantagenet and the House of Valois that started in 1337

(e) Novel by Gabriel Garcia Marquez set in the fictional town of Macondo, Colombia

(f) Administrative region of Buckinghamshire referenced when resigning from the House of Commons

(g) Period between Napoleon's return from exile and the second restoration of King Louis XVIII

(h) Russian TV game show that premiered in 1995

(i) National park in the Philippines founded in 1940

(j) Post-hardcore band from Aldershot

38. Fiat Lux

If a writing implement is driven to poverty,

and we become money-lenders,

what does an American coin become?

DECODING ZIMMERMANN

It took nearly a hundred years for the full story to emerge
of how UK Signals Intelligence helped bring the US into
the First World War on the Allied side.

On 19 January 1917 the Germans decided to carry out
unrestricted submarine warfare in the Atlantic Ocean.
The Allied blockade of Germany was biting hard and the
North Atlantic was the route by which supplies from
the US were reaching the UK and France. It was clear to
Zimmermann, the German Foreign Minister, that sinking
American merchant ships might push the still-neutral
Americans into declaring war on Germany, so he proposed
to the Mexican government that they should enter the war
on Germany's side, offering them Texas, New Mexico and
Arizona as a territorial prize for acting as a distraction
to America by causing problems on her Southern border.

The telegram containing this offer was sent to the German
embassy in Washington for onward transmission to the
German legation in Mexico City, but it was intercepted and,
unknown to the Germans, was decoded by the Admiralty's
cryptanalysts in Room 40. The complex system used by the
Germans to send diplomatic messages had become known to
the British, who were keen to exploit it. Room 40's head,

Admiral 'Blinker' Hall, realized that the German threat to give three of the United States to Mexico would help steer American public opinion in favour of war with Germany, but he faced a dilemma which has been constant in Sigint from its inception to the present day: how could he balance the use and protection of this sensitive intelligence? There is no point in spending large amounts of money to produce secret intelligence on an adversary's intentions if the intelligence can't be used. But equally, there is no point in using a piece of intelligence if doing so allows the adversary to realize that his codes are insecure and change them for something better. The intelligence has to be hidden behind a plausible alternative source, and that can take some time.

In any case, Hall first had to persuade the US that this message was real, and not a British forgery. He did this by asking the US embassy in London to obtain a copy of the original enciphered telegram from the US cable company which had delivered it to the German embassy in Washington and then letting an American official use Room 40's reconstruction of the German codebook to decrypt it himself. This convinced the US government that the telegram was authentic: all that remained was to work out how to make it public. On 28 February the story was leaked to the US press which claimed that a US journalist had obtained the message in Mexico City. To add to the confusion, Hall planted a story in the British press asking why American journalists rather than British intelligence had discovered this German plan.

The Germans couldn't try to present the telegram as a forgery, as Zimmermann himself said publicly that it was authentic on the day the story broke in the German press. A subsequent German investigation concluded not that the telegram had been intercepted and decoded, but that poor handling of classified information by the German legation in Mexico was to blame.

The sinking of American-flagged vessels would probably have brought the Americans into the war eventually, but publication of the Zimmermann telegram was decisive in persuading American public opinion that Germany had no respect for the territorial integrity of the United States. On 6 April 1917 the US Senate declared war on Germany.

39. One-part code

In a one-part code, the code groups are assigned to the letters, words or phrases that they represent in a logical order – typically, alphabetical order. A two-part code – such as that used for the Zimmermann Telegram – assigns the code groups in a random order. This has the drawback that separate codebooks are needed for encryption and decryption (hence the name 'two-part code'), but it is considerably more secure.

The following message – which you may find familiar – has been encrypted using a one-part code, in which the code groups 0000 to 9999 represent 10,000 of the commonest English words in alphabetical order. Can you decrypt the message?

4372 5464 7656 5861 0000 5432 9672 2765

0813 9981 1192 4275 0000 5432

4372 5464 7909 5861 0000 9808 2764 7782

0813 8074 8244 4506 9041 7811

9972 0361 4372 5464 9144 5861 9041 1224 3626

0813 9071 3668 0735

9041 0395 5865 3755 4792 0955 4506 9041 9847

9041 0395 4792 0955 4506 9041 9847

40. Add a number

What number is missing?

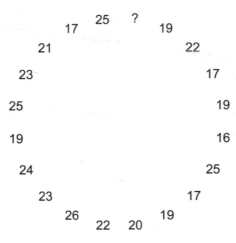

```
              25    ?
         17            19
     21                     22
   23                          17
 25                              19
 19                              16
   24                          25
     23                      17
       26              19
          22    20
```

41. Round Britain Quiz I

This style of question commemorates the *Round Britain Quiz* on BBC Radio 4: the format is one long cryptic question that has six parts to it, indicated by the letters (a) to (f). To gain full marks you should identify all six parts. This will be sufficient to answer the question.

Twelve results from an Isis-related incident on Independence Day (a), among which two short athletes in disguise (b) unusually met a day apart (c). In three of the ten others: two, five and seven symbolically convert York to Lancaster (d); 22.86cm is achieved at the end (e); and a frightening use of French is made (f). How is this?

42. Elements of a solution

Which three elements?

42.6−13.1−9.5+((20.7+59.4)/(19.3+25.2))−1.3+43.7+2.4−37.7−12.9

43. What's in a name?

Which one name could fill all these gaps?

ANNIE, ?, ?, ?, ?, BILL, VICKY, TEDDY, ?, TEDDY, ?, BETTY

44. Spot the connection

Can you see how the following words are connected?

Chuckle, Double, Euro, Pro, Stereo, Super, Tunnel

45. Divide into pairs III

Divide into pairs:

BRAIN, CRUST, EAR, FOREST, GOBLET, GOLFBALL,
GREEN, GREEN, KNIGHT, LAMB, LEATHERY, ORANGE,
PARACHUTE, PINBALL, PLUM, POCKET, POD, ROOTING,
RUBBER, SHANK, SKINHEAD, SULPHUR, WAXY, YELLOW

46. Daytime

If Groundhog Day is 1, Bastille Day is 2, Valentine's Day is 7, then
Hallowe'en is ?

47. Missing animal

(a) What animal completes this sequence: IDYHHTS, DTDWEOS,
EHAORRC, ?

(b) Answer the previous question.

48. Trivial connection

Identify:

(a) 1977 film about a faked Mars landing

(b) Father of Elizabeth II

(c) Band formed in 1981 by former members of The Specials

(d) Superhero team created in 1961 by Stan Lee and Jack Kirby

(e) Duet from *No, No, Nanette*, sung by Nanette and Tom as they
imagine their future

(f) Game in which players drop red and yellow discs into a
rectangular blue frame

(g) US TV series (1994–95) about an expedition to settle on another planet

(h) Rocket used for NASA's manned missions to the Moon

(i) Which date in July?

49. Where?

Where's WALLY?

ME	WARD	BOLD	VESPER	PREMATURE
TELEMACHUS	BLACK	STICKS	JUGGLING	GREEDY
PATIENT	PAGODA	SPONGEBOB	ALERT	SONG
LIGHTER	CHEERFUL	SAD	SCOTCH	BART
KITT	PURLOINED	UNOILED	SIDED	DATA
WORLD	BONK	RIVER	BLOBBY	ALLEGRETTO
DOGGY	HAUNTED	TOTEM	CORN	OVAL
POETIC	PROTECTION	HANGER	START	CONNECTING
VOGUE	LIFTING	AWAKE	BLESSED	STUPID
REDIPS	POGO	WATCH	MUCH	SPIN
PHOEBE	MARIO	EARLOCK	LIZARD	CONQUEROR

50. Raining

Divide into two sets:

ALLEY, BOB, BULL, HANG, HOT, POLE, THUNDER, WATCH

51. Party time

A shy ghost invited his three friends Bonnie, Nutsy and The Brain to a party. They were all busy, but each sent another in their place. The party was great and they all had a wonderful time. What was the name of the shy ghost?

52. Intriguing ways

The British have some intriguing ways. Identify the following:

(a)

(b)

(c)

(d)

(e)

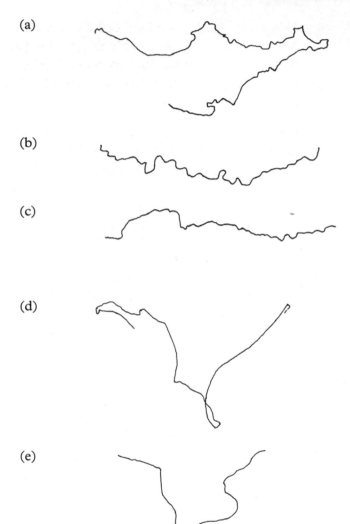

53. Registered places

The following lists are all versions of the same statement. Identify the statement and explain.

1962–1974: Middlesex, Nottinghamshire, London, Cheshire, Aberdeenshire, Manchester, Norfolk, Sussex (E), Yorkshire (N), Manchester, Yorkshire (W), West Bromwich, Staffordshire, Hertfordshire, Middlesex, Aberdeenshire, London

1974–2001: London (NE), Nottingham, London (C), Chester, Aberdeen, Manchester, Norwich, Brighton, Middlesbrough, Manchester, Leeds, Dudley, Stoke-on-Trent, Luton, London (NE), Aberdeen, London (NW)

2001–present: Manchester, Reading, Beverley, Manchester, Glasgow, Newcastle, Peterborough, Norwich, Carlisle, Newcastle, Bristol, Chelmsford, Reading, Reading, Manchester, Glasgow, Borehamwood (previously Stanmore)

54. Can't be done

This sequence can't be completed. Why is there no word in the 10th position?

MOZART, MAYBE, EXCEL, BAWDY, VENERATE, CUFFLINKS, CATGUT, BASHFUL, VARIABLE, ?, TOPKNOT, FOOLISH, CONMAN

SIGNALS INTELLIGENCE BETWEEN THE WARS (LIFE BEFORE BLETCHLEY)

The achievements of Bletchley Park were built on a foundation of the success of British Signals Intelligence between the wars.

Within a couple of weeks of the Armistice that ended the First World War on 11 November 1918, work was beginning to try to devise the shape of the UK's post-war Signals Intelligence organization. Everybody agreed that Sigint had proved its value during the war, but there were disagreements about who should be responsible for it, and how broad a remit it should have. Essentially, the Admiralty won: the new organization (to be called the Government Code and Cypher School — GC&CS) would be a civil department of the Admiralty. The head of the school, A. G. Denniston, and the majority of the staff would come from Room 40, the Naval Sigint effort, rather than the Army's effort in MI1(b). GC&CS transferred to Foreign Office control in 1922 as by that time, with no military threat to Britain and the Empire, it had only diplomatic targets and the Admiralty was facing extreme cuts. Even so, it took the appointment of Admiral Sir Hugh Sinclair

as 'C', the Chief of the Secret Intelligence Service, to
satisfy service concerns about a civilian department
of the Foreign Office being responsible for providing
intelligence support to the military.

By today's standards, GC&CS was a tiny organization: its
original establishment in 1919 was less than a hundred
people, and it had only just reached two hundred when
it moved to Bletchley Park in August 1939, shortly
before the outbreak of the Second World War. The public
mission of GC&CS was to provide advice to government
departments about the security of their codes and ciphers
and to maintain the supply of such material to them so
UK communications would remain secure. It also had a
secret Sigint directive: 'to study the methods of cipher
communications used by foreign powers'. The Sigint
mission was a resounding success, solving every book-
based code that could be collected in sufficient quantity.
There were just two exceptions – Germany, from 1919, and
the USSR, from 1927 – as they used one-time pad systems.
It must be said, though, that this owed as much to the poor
standard of security worldwide as it did to the brilliance
of cryptanalysts.

The security mission was a real responsibility, far more
than just a cover story, but there were three problems:
GC&CS could only advise and not mandate good security;
the three services, Army, Navy and Air Force, were
expressly excluded from having to ask GC&CS for advice
(though in practice they were interested in hearing
GC&CS's ideas); and the advice received from the GC&CS
seniors was not particularly good.

After its formation, the main focus of GC&CS Sigint was the Soviet Union. When the first Soviet delegation came to the UK in 1920 to negotiate recognition of the USSR by Lloyd George's government, its telegrams were all decrypted and read. The Soviet Union did not pose a military threat; the threat came from its support for the Communist Party and its attempts to create revolutionary conditions among the working class in the UK. Things came to a head when, a year after the 1926 General Strike, the Government sent police into the offices of the All-Russian Co-operative Society (ARCOS) to find evidence of Soviet interference in the UK. Forewarned, ARCOS officials had destroyed incriminating material, which meant that the government decided to publish decrypted Soviet telegrams to show why it had been thought necessary to raid the ARCOS offices.

The care taken to hide from the Germans in 1917 that the text of the Zimmermann Telegram had been obtained by cryptanalysis — they believed that theft or subterfuge lay behind its publication in the US press — was not even attempted in 1927, and the Russians, realizing that their diplomatic codes were being read, changed them. This is still the most flagrant British example of the squandering of a Sigint source for short-term advantage.

Although Soviet diplomatic messages were no longer readable, all was not lost. The Communist International or Comintern, a Soviet-controlled organization that officially linked the world's communist parties but which secretly provided a structure for Soviet espionage and subversion, continued to use codes that could be broken and read. Work against Comintern codes was led by John Tiltman whose career in UK and US Sigint lasted from

1920 to 1980. Tiltman had learned about cryptanalysis
from Ernst Fetterlein, a Tsarist cryptanalyst who had
escaped from Russia after the 1917 revolution and who was
recruited into the new GC&CS. The idiosyncratic Russian,
who had a low threshold for suffering the work of people
less competent than himself, hit it off with the decorated
infantry officer who discovered, out of the blue, a gift
for cryptanalysis. Tiltman spent the years 1921 to 1929
in India, but on his return headed both the Military
Section at GC&CS, which trained army officers in basic
cryptanalysis, and the work, codenamed MASK, which
monitored illicit Soviet traffic. MASK uncovered
an international network of Soviet activity in Europe
and Asia.

GC&CS's success against diplomatic codes was maintained
through the 1930s. But German and Russian traffic was
unreadable, and little progress was being made against
the encryption systems being used by the German military,
with whom the UK would be at war before the end of the
decade. The GC&CS that moved to Bletchley was living on
its reputation, and was staffed, with a few exceptions, by
people whose best years were behind them. Luckily, a new
stream of talent was about to join.

55. Polar trek

The following puzzle was in the arithmetic section of the GC&CS Entrance Test from 1926:

In a polar expedition a base A has been formed near to the Pole P and the final dash is to be made in three equal stages with intermediate dumps at B and C.

In organizing the polar party the supplies of food and stores are divided up into rations, each of which is sufficient to support one man for one stage. One man is able to carry three rations. The party is to start out from A fully loaded; at B it is to divide into two parts. One part is to take over rations from the other and proceed to C while the other part is to return at once to A with one ration per man after having made a dump of the remaining rations at B for the members of the first part to pick up upon their return. At C this proceeding is to be repeated and one man only is to proceed to P. This last man is to carry two rations for his support in journeying to P and back to C whence he returns to A by aid of the dumps left by his comrades. On this basis find the minimum number of men in the polar party which is to set out from A.

56. A festive answer

Suggest a musical follow-up to

ATMOSPHERIC

BLASPHEMING

CENTRIFUGAL

DOCUMENTARY

EXCULPATION

FORECASTING

57. Lobotomised a mule

(a) Fog and radar

(b) Ulster naval hoax

(c) A bad scooter

(d) Piranhas find nets

(e) A basic Irish summit

(f) Salty tattooer

(g) Accrediting a sponsor

(h) Tofu paint

(i) A transparent idiom

(j) Peanut cereals

58. Book

Which literary work can be represented by

$$\frac{S}{S} + \frac{L}{S}$$

59. Which?

The following list of 55 words can be divided into 10 sets, all of different sizes. Put another way, there is one set of 10, one set of 9, etc. Which word is in the set of one?

BANTAM	BARN	BRIG	CABER	CHURCH
CONFIRM	COR	COURT	CRUISER	DRAG
DRIFT	DRUGGED	DUNCE	EL	EMINENT
ERR	FARM	FEATHER	FISH	FLY
GESTURE	GRAVE	HAIR	HEAVY	HEIST
HINT	INTER	JUNK	LIGHT	LOFTY
MIDDLE	MO	NATIONAL	NOM	NORD
NOTICE	ONE	OP	PORTENT	QUICKS
RAISED	REM	RIB	ROYAL	SCHOOL
SEINE	SHAKESPEARE	SHIP	SON	TALL
TEMPT	TIRED	TRACT	VINE	WELTER

60. Balance the scales

One of each 1–3kg weight has been placed to balance these scales:

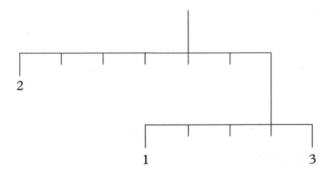

(a) Can you do the same for one of each 1–6kg weight?

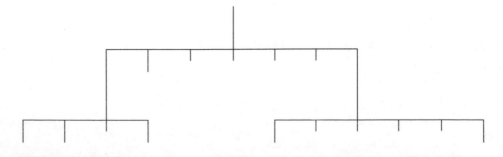

(b) 1–10kg weights?

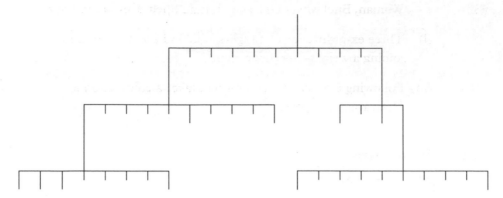

61. All the Right Movies

(a) A Lancashire boot-maker discovers there's only one type of ice-cream available.

(b) A celebrated jeweller travels to Newcastle to avenge his brother's death.

(c) A seaside resort is menaced by small cloaked figures.

(d) An Indian musician joins the rebellion against imperial rule.

(e) A protest is triggered at a military academy when a group of cadets go for Spanish food.

(f) The members of a venerable Scottish golf club are abducted into a digital world.

(g) A detective from Detroit opens a convenience store in LA.

(h) On an alien planet a man in a conical hat leads a rebellion of the indigenous people.

(i) An Egyptologist finds an ancient device for making houses child-safe.

(j) The leader of a group of Conservatives is threatened by an upstart new arrival.

(k) After being exposed to cosmic radiation, Mr Rye, Wholemeal Woman, Buckwheat and The Human Spelt team up to fight evil.

(l) Three ex-girlfriends of a high-school lothario trick him into joining a weight-loss programme.

(m) Following a botched job, a hitman takes a contract on a Hungarian composer.

62. Prime time

24	→	31
60	→	211
144	→	42
500	→	203
360	→	?

63. Rhyme time

Fill in the gaps:

?ce ?ass ?ase ?ace

64. Sums

(a) Aus* − D*on × I*sco = ?

(b)
$$\frac{\text{battle} + \text{conquest}}{\text{beneath} + \text{dawn} - \text{rise}} = ?$$

65. Elimi-nation

Which country is missing?

INCH, TIARAS, AIRY, AVAIL, MANGEY, ALONG, RAINED, CANDLE, MARINE

66. Football logic puzzle II

In the following table every pair of teams has played each other once. The table is ordered in the usual way (3 for a win, 1 for a draw) with goal difference, goals for and fair play record used as tiebreakers. Using arithmetic and logic, fill in all the gaps.

Team	Points	Goals For	Goals Against	Goal Difference
United		9	7	
Wanderers		4	3	
Rovers			0	
City				−3

Wanderers	*v*		City
United	*v*		Rovers
Rovers	*v*		City
United	*v*	3	Wanderers
Rovers	*v*		Wanderers
City	*v*		United

67. Complete . . .

Complete the sequence:

OC, SL, L, C, M, LC, C, B, MG, LG, G, ?

68. Clock I

There is a clock in the puzzle archive – a normal clock with the numerals 1 to 12 around the clockface. However, recently someone – who presumably wants to get revenge on the setters by giving them some puzzles to solve – has taken to sticking words or numbers over these numerals and moving the hands to show the time indicated by these.

The setters have solved these puzzles, naturally, and have therefore established that the words or numbers in each case form a sequence going around the clockface. However, as soon as we have solved a sequence and removed the stuck-on words or numbers from our clock, another sequence appears the next day. Below are some of these sequences. In each case, what time showed on the clock?

(a) Gin, Cheese, Eric, Snakes, Chapter, Accident, Stan, Arts, Mapp, Roland, Noughts, Lock

(b) TACKLE, SWITCH, ENERGY, UNCLIP, NATIVE, MYSELF, FIXING, BISQUE, JOVIAL, GATEAU, HOTPOT, ZODIAC

(c) 52, 7, 34, 27, 60, 16, 4, 9, 8, 75, 28, 10

(d) Morocco, Poland, Italy, Armenia, Oman, Guinea, Hungary, Japan, Estonia, Kenya, Portugal, Macedonia

(e) POINT, INDIA, VNECK, TASTE, FIGHT, AWASH, VOICE, ENUFF, FARSI, INOFF, SHOOT, EARTH

(f) CSC, NOT, XMUTI, KICW, XNOY, AFY, QWGBG, DOZKW, HFQV, IXD, HKQIIY, TSXKXY

69. Turning Japanese

How might you turn?

Guinea – Bolivia

Italy – Hungary

France – Netherlands

Peru – Austria

Ireland – India

Japan – Japan

70. Missing characters

Identify the following books from their opening lines. We've included only letters from a subset of the alphabet (a different subset in each case).

(a) T-e a-c-em--t p-c-e- -p a b--- t-at --me-ne -n t-e ca-a-an -a-b----t. -eaf-n- t------ t-e pa-e-, -e f--n- a -t--y ab--t Na-c-----.

(b) V-----m, m- l-v- --u l--g --m-. -ll d--, -ll --g--, m- l-v- --u l--g --m-.

(c)   I- w-- ---- -v--i-- w--- K ---iv-d. --- vi----- --y d--- i- ---w.

(d)   A--o- -o-- -'a--i-ai- -a- à -o--i-. I- a--u-a. -o- -a- -a--uai-
      -i-ui- -i---.

(e)   T-- ---z- -- --- ----g p-p- -- ----t ---- t-- --d ---'- ----, --t -- --pp-d
      -i- ---d- -----d t-- g---- -- --t ----t t-- -- i- t- ---- t---.

(f)   H---r- R--r- l---h--. H- st--- n---- -t th- ---- -- - cl---. Th- l--- l-y
      --r b-l-- h--.

(g)   A -o--e -oo- a --ro-- --ro--- --e -ee- -ar- -oo-. A -o- -a- --e
      -o--e a-- --e -o--e -oo-e- -oo-.

(h)   I- - h--- i- -h- ---u-d -h--- -i--d - h---i-.

71.   A bag of words

(a)   What property do these words share?
      BAD, BALL, DAN, MASS, PAT, TAN

(b)   What connects the following?
      FAND, STECK, TIRN, HOM, TULL.

SOLVING ENIGMA — PART ONE

Enigma, an electromechanical rotor cipher machine, was
introduced in the 1920s for use by businesses to keep
commercial communications secret. Its novel technical
characteristics made it the cutting-edge technology of
its day, which is why UK Signals Intelligence set out to
understand it long before it was used by German military
forces in wartime.

In the spring of 1926 a German engineer called Arthur
Scherbius launched an improved version of his Enigma
enciphering machine on to the market. Scherbius
believed that his machine would enable businesses to
send completely secure messages through the public
telegraph system. Known to us as Enigma D, it was
simpler than the subsequent military versions of Enigma.
(see Part Two). Recognizing the potential of this new
technology, the deputy head of GC&CS, 'Jumbo' Travis,
went to Berlin in November 1926 and bought one for £30
(equivalent to about £1,200 in 2018). He brought it back
to the UK and asked the most junior cryptanalyst in
GC&CS, Hugh Foss, to produce a diagnosis of the machine:
how did it encipher messages, and how might a message
enciphered on it be deciphered?

Enigma has a keyboard like a typewriter but when an operator presses a key, instead of printing the letter on a piece of paper its encrypted equivalent lights up on the lampboard. Foss discovered that Enigma encrypted messages by passing a current through a series of rotors and a reflector that transposed the input letter a number of times. Given that the three rotors of an Enigma D could be placed in any order, and that the reflector could be set at any one of twenty-six positions, it could produce 2,741,856 different substitution alphabets.

Scherbius was convinced that a machine capable of so many permutations was unbreakable, but Foss discovered that there was a fundamental flaw in the design of the machine. Enigma machines are reciprocal: for any given setting of the rotors, if P produces Q, Q produces P. And that in turn means that a letter can never be encrypted as itself. Foss thought through the implications of this: the chance of two separate strings of letters not containing any matches becomes vanishingly small once you reach about forty letters, so by comparing a long encrypted message with a long piece of unencrypted text an analyst would expect the message to contain, finding the message in which there were no matches at all could lead you to the message you were looking for.

The early cryptanalysts used a piece of schoolboy slang to describe this piece of predicted plaintext: they called it a 'crib', the word used for the translations of Greek and Latin texts schoolboys would smuggle into their Classics classes. Most of the early cryptanalysts recruited into Room 40 and MI1(b) during the First World War had received a public-school education with a strong focus on Classics. One of them, Dilly Knox, had become a palaeologist and spent much of his life working on the *Mimiambi*, fragments of Greek verse that needed to be reconstructed from scraps of papyrus that had been found by archaeologists in the late nineteenth century. The skills he needed for this task went beyond linguistic ability, which was a given: he needed an understanding of Greek literature and of the mindset of a classical author if he was ever going to be able to piece together the fragments to produce a definitive text. These proved to be precisely the skills that were needed in the first cryptanalysts who were brought together in the first months of the First World War. Knox was recruited to Room 40 early in 1915 and was employed as a cryptanalyst until his early death from cancer in 1943, aged fifty-eight.

Knox had been fascinated by the machine Travis brought back from Germany and Foss's diagnosis of it, and, building on Foss's first ideas, had developed a theoretical cryptanalytic attack on Enigma, designing a set of cardboard rods which would allow him to mimic the action of the rotors inside Enigma.

Early in the Spanish Civil War, Hitler sent a military force, the Condor Legion, to Spain to fight for Franco's forces. They brought with them ten Enigma K machines to secure the communications coordinating activity between German, Italian and Spanish nationalist forces. The Enigma K was a rewired version of the D variant, and Knox's rods were effective in breaking into this network in April 1937. This success was qualified, however, by the fact that for their own use the Germans were using a different and much more complex variant of Enigma, which would not yield to rods. Instead, and for the first time, electromechanical encryption would have to be defeated by electromechanical decryption.

72. Enigma Variations I

Fill in the gaps such that once complete each line will in turn lead somehow to the letters E, N, I, G, M, A.

(a) Thor's son is _ _ _ _ _ .

(b) The comic book company is _ _ _ _ _ .

(c) Drake's sister is _ _ _ _ _ .

(d) The pine tree state is _ _ _ _ _ .

(e) The Rolling Stones single is _ _ _ _ _ .

(f) The Latin element is _ _ _ _ _ .

73. A question for Mum and Dad

What connects the following?

A flightless bird, a bloodsucking fly, a softly spoken utterance, a French novella, steamed semolina and some exceptionally able people.

74. Round Britain Quiz II

This style of question commemorates *Round Britain Quiz* on BBC Radio 4: the format is one long cryptic question which has six parts to it, indicated by numbers in parentheses. To gain full marks you should identify all six parts. This will be sufficient to answer the question.

What might be said at dusk by others (2); personally inexplicable by unknowns (6); sadness by the rivers (8); geometry by US prisoners (10); a nostalgic question by seemingly crooked people (12) appear on what sounds like a set of posters (1–12, later 1–14) that followed a mad boy. How is this?

75. The element of surprise

Here's a nice straightforward question. Or at least it would have been if one of the setters hadn't knocked over the ever-unstable kwiz koffee kup and obscured the ends of the instructions. So now it's just a nice question :-)

(1) Begin with a list of the 118 elements. Eliminate the 83 which

(2) Eliminate the 4 elements which begin with the letter

(3) Eliminate the 3 elements which end with the letter

(4) Eliminate all the elements which begin with the letter

(5) Eliminate all the elements which begin with the letter

(6) The remaining elements can be divided into 2 sets of equal size based on whether they are of odd or even length. Eliminate the half which are of

(7) Eliminate the 3 elements which are of length

(8) Eliminate the 5 elements whose names do not contain the letter

(9) Eliminate the element of length

(10) Eliminate the element whose final letter has a Scrabble value of greater than

Which element remains? And which element was eliminated in step 10?

76. Sounds expensive

I blew my savings (and more) on a world tour. In South Africa I marvelled at the SAND. In Japan I discovered ZEN. Did I find a CRAB, a SEAL or a SNAKE in Brazil?

77. **Year anagrams**

In each part there are three events in world history given in chronological order. The years that the three events occurred in are anagrams of each other. What are the years?

For example:

1294: Kublai Khan dies

1492: Christopher Columbus lands in the New World

1942: The Battle of Stalingrad starts

(a) The future Henry II marries Eleanor of Aquitaine; Magna Carta is signed; Martin Luther is excommunicated

(b) Fall of Rouen to Henry V; Henry VIII is born; Pearl Harbor is attacked by the Japanese

(c) Emma, mother of Edward the Confessor, dies at the age of 67; Ferdinand Magellan becomes the first European to sail from the Atlantic Ocean to the Pacific Ocean; Sepp Blatter resigns as FIFA President

(d) Napoleon Bonaparte invades Egypt; William McKinley is sworn in as 25ᵗʰ President of the USA; Andy Murray is born

(e) Richard Cromwell resigns as Lord Protector of England; The USSR invades Hungary to quash the uprising there; Winston Churchill dies

(f) The current St Paul's Cathedral consecrated for use; Captain James Cook lands in New Zealand; The Six Day War (Arab–Israeli)

(g) Christopher Columbus dies; The Gunpowder Plot; King William III of England is born

(h) Queen Victoria is born; American composer Cole Porter is born; Iran releases 52 Americans after 444 days in captivity following the Iranian revolution

(i) Dick Turpin is hanged; George Washington is sworn in as President for the second time; The UK joins the European Economic Community

(j) Saladin captures Damascus; Battle of Tewkesbury (part of the Wars of the Roses); George I becomes King of England

78. Books

Identify the following books from the word lengths of their titles, authors and opening lines. All punctuation has been retained apart from apostrophes, which have not been counted.

(a) 10 by 1 1 5   3 4 3 5 3 5 3 7 4 4.

(b) 3 6 2 3 4 7 by 1 1 5   5 3 1 3 6 7 8 6, 3 2 6 8 2.

(c) 6 3 6 by 1 1 1 6   3 3 2 4 2 8, 3 3 6 2 3 5 3
                            3 4 3 4.

(d) 1 4 4 1 4 by 1 1 7   '3 7 3 2 8 2 2 2,' 4 4 8,
                            '2 8 2 3.'

(e) 3 5 6 by 1 5 10   2 2 7 3 4 10 5 2 6 4 2 4 6
                            4 3 4 7 4 2 2 4 4 5.

(f) 3 6 7 by 1 1 7   5 11 3 3 4 2 2 3 2 6.

(g) 3 9 6 2 2 by 1 5 4   7 5 2 3 5 2 3 5 6 8, 4 5 5,
                            3 3 1 6, 3 4 2, 3 3 3 7 4.

(h) 5 2, 6 by 1 1 9   3, 8 4 8 2 3 6 – 2 3, 3 4 –
                            3 2 2 5?

(i) 6 1 2 2 5 by 1 1 6   3 7 2 7.

(j) 3 5 2 3 5 by 1 1 5   2 7, 10 3 7 2 3 5 4 3 8 9,
                            5 3 4 2 3 4 2 3 3 7, 10 3 9.

(k) 1 4 2 3 5 by 1 1 7   3 5 2 4 2 2; 3 3 3 7, 3 5
                            10 2 6 7, 4 2 5 2 2.

(l) 2 5 7 by 1 8 7   3 3 5 4 3 4.

The GCHQ Centenary Trail

Welcome to the Centenary Trail. This is a connected series of puzzles throughout the book which between them should thoroughly test your puzzle-solving mettle!

Here's how it works: On the next two pages are a pair of puzzles, the first of which is A VERY STRAIGHTFORWARD QUESTION. Solving these will lead to a year, and a description of a part of this book. When you turn to this part of the book you will find another puzzle. The solution to this will also be a year, together with the title of a puzzle in this pictorial inset. The solution to the named puzzle will be a year, plus a description of another part of the book, to which you should then turn. And so on. You will find a piece of ciphertext below seven puzzles in this inset. When the associated year is even you should treat the ciphertext as having been reversed.

As you make progress you should find yourself compiling a list of years and ciphertexts, and then you just need to work out what to do next!

The puzzles vary in difficulty, but we hope you find them fun. There is no prize this time sadly — just well-deserved satisfaction. Good luck!

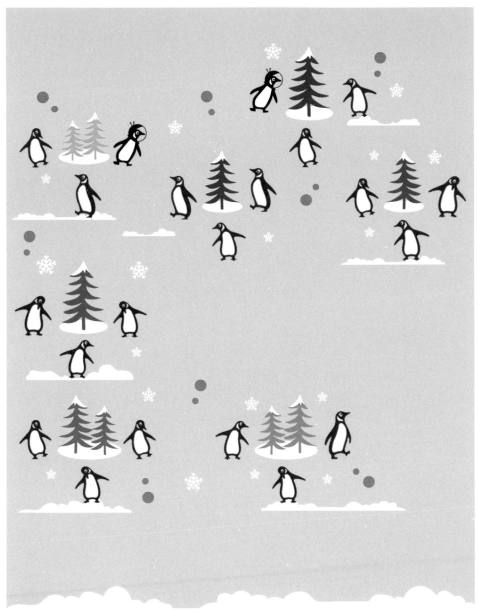

Grid

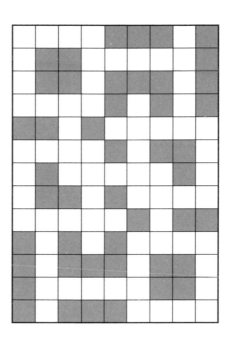

Fit the following words into the grid. There are three words missing – what are they?

All letters of the alphabet appear in the finished grid.

ABLE   BHAJI

AXIS   CIVIL

CALF   EQUAL

EVEN   IRISH

LAWN   MANIC

PAIR   QUICK

PILL   SITAR

TIDY   YACHT

TUBA

TURF

ZEST

ZINC

Penguin
puzzle

QKTTXXSIBXNFQ

Alphabet St.

Finish

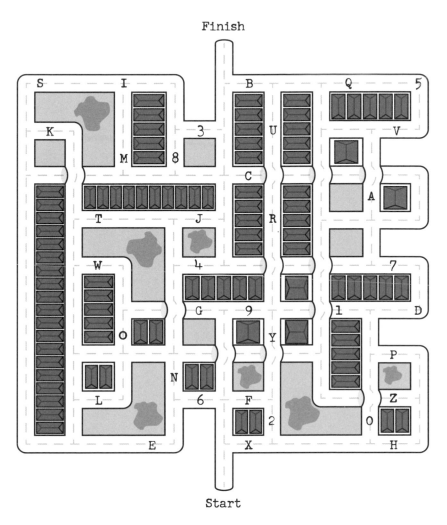

Start

Find the shortest route in the maze from the start to the finish.
When going over or under a bridge, you may only go straight on.
At all other junctions, and at corners, you may NOT turn right.

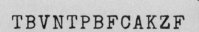

TBVNTPBFCAKZF

Obfuscated
code

```
function  encrypt(a) {

        var A = "nymphsblitzquIckvexdwarfjog.";
        var b = A, n = 28, i = 0, z = 0;
        var c = "";

        while (i < a.length) {
                x = A.indexOf(a.charAt(i));
                y = A.indexOf(b.charAt(i % b.length));
                z = (x + y) % n;
                c = c + A.charAt(z);
                if (x + 1 == n) {
                        b = c;
                }
                i = i + 1;
        }
        return c;
}

>> console.log(encrypt(SECRET));
tsdmueyuvrxIedqqfmdqweIyaaxtiyzrujqezxqdawgotw
```

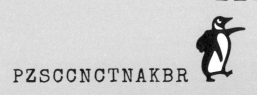

PZSCCNCTNAKBR

Regex crossword

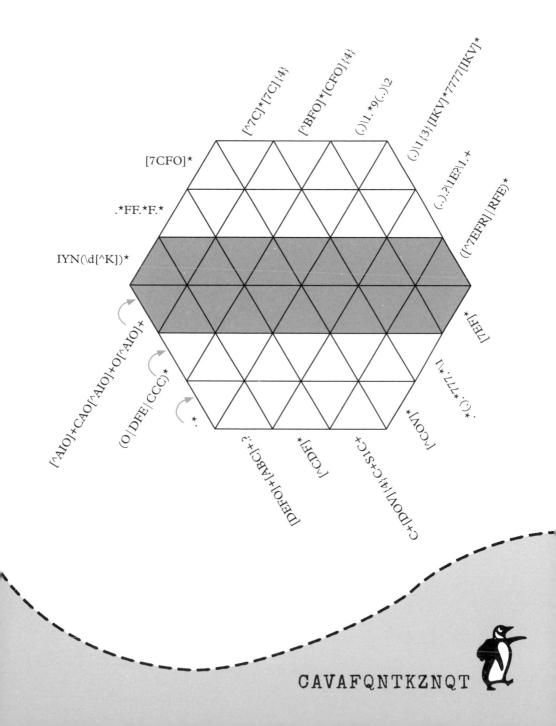

Games

It is probably not surprising to hear that many GCHQ puzzlers also enjoy games. As you might expect, those that enjoy games enjoy puzzles to do with games. Have a go at these, you can look up the rules online if you need to.

1. Some Games

(1) Chess: I was playing monochromatic chess with a friend. In this game a piece has to move to a square the same colour as the square it started on. Part way through, a piece was knocked off the board leaving it looking like this:

Which square was it knocked from?

(2) Omaha Hold'em:

Your hand is   The table is

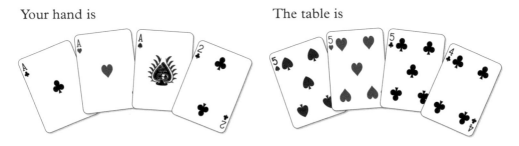

The river card gives you a cinch (a hand that cannot be beaten nor equalled).

What is the river card?

(3) Mornington Crescent: While browsing the tfl.gov.uk website, my friend and I invented a blitz-variant of the classic game that is guaranteed to finish within 11 moves provided no station is repeated. The Bakerloo, District and Jubilee lines are completely removed from consideration.

Here is a sample game:

- Chancery Lane
- Tottenham Court Road
- Euston Square
- Russell Square
- Mornington Crescent.

What name did we give to our variant?

(4) Bridge:

You hold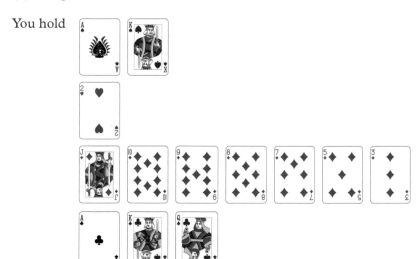

Your partner opens 5♥ which you interpret as a Grand Slam force in hearts.

What is your bid?

(1) Battleships: It had been a confusing game of Battleships and I wish I had remembered the order of my guesses. The ship sizes were 5, 4, 3, 3 and 2. One more shot would finish me unless I could win this turn. Where should I play?

(O = 'Miss', X = 'Hit', S = 'Sank')

	A	B	C	D	E	F	G	H	I	J
10			O							
9		O								
8							O			
7			O		O					
6		O	X	S	X	S				
5		O	X	X	X	X		O		
4				S	X	X	X			
3				X	S	X	X	O		
2							O			
1	O							O		

(2) Hangman: Below is the closest game of hangman that I've played for a while. Fortunately, there was only one possible answer after my last guess. Unfortunately, I can't remember my first guess which can't be read due to the coffee stain.

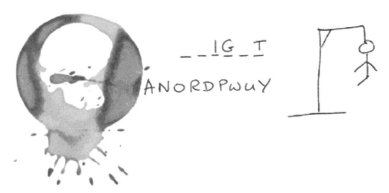

_ _ _ I G _ T

A N O R D P W U Y

What was the word?

(3) Cribbage (2 player): I played an unusual hand of Cribbage last night. Given the cards in my hand, my opponent's hand and the box, no cut could improve the score of any of them. However, if either of us had discarded differently an improvement would have been possible. My opponent played both the first and the last card. What was my hand?

3. Su-Donut-Q

We never met a sudoku we couldn't look up to.

There are three possible solutions to this sudoku.
Choose one that you think appropriate.

	A	B	C	D	E	F	G	H	I
9			9	6	5	4	7		
8		6						9	
7	5			2	9	3			8
6	4		6				8		9
5	8		7		4		3		1
4	9		3				6		2
3	1			5	6	2			7
2		3						8	
1			5	8	3	7	1		

Resistance is futile

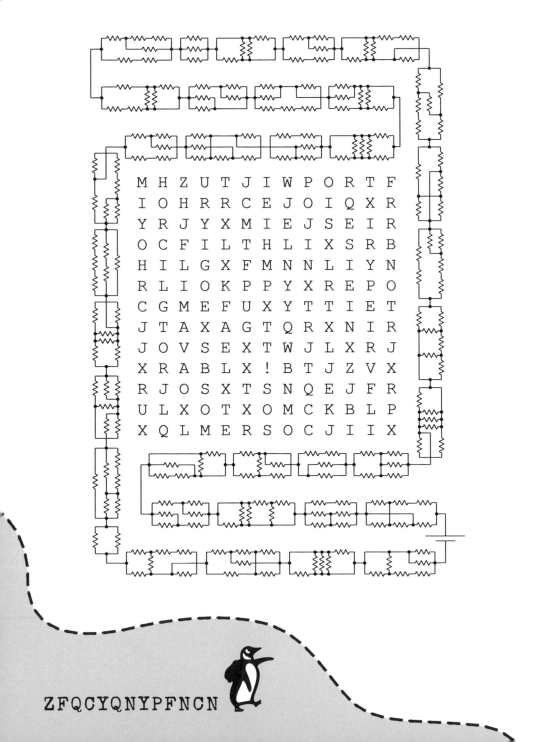

```
M H Z U T J I W P O R T F
I O H R R C E J O I Q X R
Y R J Y X M I E J S E I R
O C F I L T H L I X S R B
H I L G X F M N N L I Y N
R L I O K P P Y X R E P O
C G M E F U X Y T T I E T
J T A X A G T Q R X N I R
J O V S E X T W J L X R J
X R A B L X ! B T J Z V X
R J O S X T S N Q E J F R
U L X O T X O M C K B L P
X Q L M E R S O C J I I X
```

ZFQCYQNYPFNCN

A Century

1919 NAB the FFM of P	1947 DB is B in B
1920 UW is B at WA	1948 NHS is E
1921 AEW the NP for P	1949 MZP the CPR of C
1922 HCD the T of T	1950 S of KW
1923 KAB the FP of the R of T	1951 F of B is O by KG the S
1924 RMB the FBLPM	1952 QE the SA to the T
1925 FD of T by JLB	1953 D of JS
1926 UKGSB on F of M	1954 RBR the FMM
1927 RCG to BBC	1955 RE is the LWH in the UK
1928 P of 'TTO' by BB and KW	1956 FESC is H in S
1929 MLK is B in AG	1957 EEC is F by the T of R
1930 P is D by CT	1958 EMUPD in the MAD
1931 FP of the HC	1959 P in CS by FC
1932 K of CALJ	1960 OT of 'LCL' by DHL
1933 AHBC of G	1961 YG is FM in S
1934 BP and CB are K in an A	1962 CMC
1935 FV of M is S by PB	1963 JFK is A
1936 A of KE the E to MWS	1964 NM is S to LI
1937 O of the GGB in SF	1965 MX is A
1938 FA of S in DCC	1966 EW the WC
1939 SHSB is D	1967 TBR 'SPLHCB'
1940 O for BOSW by 'TWOO'	1968 MLK is A
1941 PH is A by J	1969 PC is I as P of W at CC
1942 B of SH, the A of 'ABHOT'	1970 S of UT
1943 DR on the RV	1971 DC is I in the UK
1944 DDL in N	1972 O of SJ at GH
1945 AB are D on H and N	1973 UKIVAT
1946 EHD the PR of A	1974 AS is E from the SU

1975	JC the FBK of S	2003	LF of C
1976	HWR as PM of B	2004	MZCF
1977	VWWWLSF	2005	AMB the FFC of G
1978	ACRA off B	2006	P is D from P to DP
1979	RL at TMI	2007	F iP is R
1980	S at IE in L	2008	C of the LHC
1981	FLM is R	2009	FBB is E by SN
1982	FW between UK and A	2010	DHODPE in the G of M
1983	S is K in CK	2011	W of PW and CM
1984	IRAB of the GH in B	2012	OGH in L
1985	FE of 'E' is B	2013	R of PB the S
1986	C of the MTF	2014	MAFS is SD over U
1987	H of FEFC	2015	QE the SB the LRBM
1988	CDW the ESC	2016	P of 'TGCHQPB'
1989	TBLI the WWW	2017	RM is D as P of Z
1990	D against the CC	2018	W of PH and MM
1991	BYB the FP of RF	2019	GCHQC
1992	EUE by MT		
1993	NPPW by NM		
1994	CTO between E and F		
1995	FLTP by AW		
1996	R of 'TL' by B and S and LS		
1997	P of 'HPATPS'		
1998	GFAENIA		
1999	W of VA and DB		
2000	FE of 'DTE' on N		
2001	O of F and MD in B		
2002	BPW by YM for 'LOP'		

Below is a picture of the trophy awarded to the winner of the competition in the first GCHQ Puzzle Book. It has inscriptions on each of its 3 sides, and these inscriptions are shown on the next page. Unsurprisingly they comprise a puzzle!

L
I
B
E
R
A
L

79.   **Film outlines**

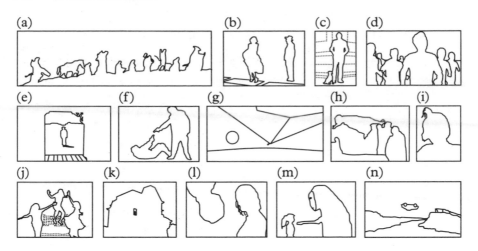

80.   **On the turn**

IF YOU CAN'T RESIST A RIDDLE

THEN LET THIS WHIRL AROUND YOUR MIND

YOU MAY CANCEL OUT MY SYMBOLS

BUT I'LL STILL BE LEFT BEHIND

WHO : W/V = I?

81.   **Pairs**

Pair each word in the first list with an element from the second list to find where they belong.

List 1:   AGUE, BERN, HARE, LION, LOON, MAMA, MINK, MONO, QUIT, SOFA

List 2:   ACTINIUM, ANTIMONY, IODINE, LITHIUM, NEODYMIUM, OXYGEN, PRASEODYMIUM, RADIUM, SODIUM, SULPHUR

82. Football riddle

Last night I was watching a football game and I noticed that one of the teams had a nickname with an interesting property.

The first few letters of their nickname was an anagram of a planet.

The last few letters were an anagram of a different planet.

What was the football club?

83. Sums

(a)
$$\frac{\text{do admin} + \text{red flaps}}{\text{spy mug}} - \text{cat lice} = ?$$

(b) Morning + The Hen + (Mercury − Evening) / Noon = ?

(c) $(\text{Rise} - \text{Revenge})^{\sqrt{(\text{Fate} - \text{Power})}} = ?$

(d)
$$\frac{\text{Boswells}}{\text{Dursleys}} \times \frac{\text{Mr Benn} - \text{Winfield/Powers/Flanders}}{\text{Paddington} - \text{Sirius}} = ?$$

84. Not a letter

Fill in the blank ... with a number!

$$
\begin{array}{c}
? \\
\text{W} \quad \text{E} \\
\text{S}
\end{array}
$$

85. Have a positive attitude

A cyclic shift is a simple cipher where you shift each letter of the alphabet on by a fixed amount, wrapping back around to the beginning of the alphabet if you get to the end. For example, with a shift of 3 A→D, B→E, C→F, ... Z→C.

Can you name a three-letter English word which is a cyclic shift of its translation into French?

86. Pink zoom bearer

(a) Mulish trainee

(b) Shunted data

(c) New maniac

(d) Career type

(e) Broke in

(f) Loud chalet

(g) Thermal inlay

(h) Horrid music

(i) Drinkers clash

(j) Smashed urinal

87. Missing animal

Fill in the gap:

Snake, Penguin, Dog, Ant, ?, Crab

88. The Great Puzzle Bake Off

The Great Puzzle Bake Off got off to a slightly shaky start. An error with capitalization when advertising for entrants meant that Bakers, rather than bakers, were asked to apply. Nonetheless, 12 Bakers did take part: Anita, Cheryl, Colin, Danny, George, Ginger, Josephine, Kenneth, Matt, Richard, Stanley and Tom.

The format was the same as usual. Each round involved three bakes – the signature bake, the technical challenge, and the showstopper – and at the end of each round the judges decided who would be eliminated and who would be honoured as 'Star Baker'. After 9 rounds the final was held between the 3 remaining entrants.

Below is a summary of the event showing the challenges and the results. Unfortunately someone knocked over a glass of cooking sherry, which makes some of the summary unreadable.

Round	Theme	Signature	Technical	Showstopper	Star	Eliminated
1	Breads	Focaccia	Italian grissini	Soda bread	Anita	Kenneth
2	Biscuits	Caramel shortbreads	Nice biscuits	Biscotti	Tom	Danny
3	Pastry	Charlotte royale	Flaugnardes	Individual suet puddings	Cheryl	Tom
4	Tarts	Kiwi fruit tart	Tarte tatin	Blue-cheese quiche	Anita	Matt
5	Pies	Spicy lamb pasty	Raised game pie	Double-crusted redcurrant pie	Colin	George
6	Puddings	Eggnog pudding	Lemon soufflé	Syrup sponge pudding	?	?
7	Special diets	Low-fat choux puffs	Vegan sherry trifle	Plaited wheat-free loaf	?	?
8	Patisserie	Éclair au chocolat	Espresso mousse	Vanilla macarons	?	?
9	Cakes	Chocolate layer cake	Cuba rum baba	Upside-down cake	?	?
Final	European	Spanische windtorte	Sfogliatelle	Hazelnut dacquoise	?	–

(a) Who was eliminated in Round 9?

(b) Who won the competition (i.e. who was Star Baker in the final)?

(c) Who presented the award?

SOLVING ENIGMA — PART TWO

When it came to solving the military variant of Enigma,
UK Signals Intelligence looked to foreign allies for help.
Combining the efforts and the different approaches of
British, French and Polish cryptanalysts ultimately led
to success against the most widely-used encryption device
of the German armed forces.

In the latter part of the 1920s the German military had
adopted Enigma as their principal operational encryption
system. They added a plugboard to the front of the machine,
increased the number of rotors to five, from which
the three to be used would be selected, and introduced
standard operating procedures which they believed would
ensure the security of the machine. But the fundamental
weakness of Enigma — the fact that its design meant that
a letter could never be encrypted as itself (a gift to
cryptanalysts) — was never addressed.

The UK was not, of course, the only country interested
in understanding what the intentions of Nazi Germany
might be: both France and Poland understood that
decrypting German military communications was key
to understanding German plans. The French had used
traditional espionage techniques to recruit a venal

German Army signals officer who supplied them with
handbooks and cryptographic material in return for
money to fund a lavish lifestyle. The Poles recruited
a team of mathematicians to study Enigma as a
mathematical problem. The sheer brilliance of Marian
Rejewski and his colleagues, notably Jerzy Różycki and
Henryk Zygalski, gave the Poles sufficient insight into
military Enigma not just to produce intelligence from
intercepted messages but to conceive and build machines
to support Enigma cryptanalysis. But changes introduced
by the Germans to Enigma made them ineffective.

In 1938, A. G. Denniston had begun to recruit
mathematicians for service with GC&CS during wartime,
most famously Alan Turing. In July 1939, French, Polish
and British cryptanalysts met in Warsaw and shared all of
the information they had about Enigma. The information
provided by the Poles inspired the section run by British
cryptanalyst Dilly Knox, which had moved to Bletchley
Park in August 1939 with the rest of GC&CS, to try to build
on the progress the Poles had made.

Turing and others, no doubt inspired by Knox's reports,
began to think about electromechanical solutions to the
Enigma problem. After a meeting with the Poles in January
1940 in Paris — they had escaped Poland when it fell to
the Nazis — Turing began to conceive a new machine. His
bombe was not simply a development of the Polish bomba,
after which it was named: it was radically different,
working from a 'crib' rather than from assumed settings.
Once another of the new mathematician recruits, Gordon
Welchman, had seen a way of significantly improving
on Turing's design, the bombe became the workhorse of

Enigma cryptanalysis, not just mechanizing the process but industrializing it as well: by the end of the war some 30,000 people were working in UK Sigint.

The bombe itself didn't break Enigma messages: it simply reduced the number of possibilities cryptanalysts would have to explore to break them, but cryptanalysts were still necessary to exploit the results the bombe produced. There wasn't an obvious way to identify the sort of person who might have an aptitude for such work, but they were found. An unexpected consequence of the mass conscription of young men to the armed forces was that women were recruited into Bletchley Park in large numbers. Indeed, by the end of the war 76 per cent of those working at Bletchley were women, including Joan Clarke, a mathematician who worked alongside Alan Turing in Hut 8 breaking German Naval Enigma. Knox, idiosyncratic as ever, surrounded himself with young female graduates: they were known as 'Dilly's Girls' but the flippant description disguised the fact that Knox selected them from the notes written by an interviewer at the Foreign Office. He didn't want people who had got into Bletchley either because their families had influence, or because they had been conscripted: he wanted intelligent women who were looking for an aspect of war work to which they could contribute. Their best-known success came when Mavis Lever (later Batey) decrypted an Italian naval message that gave the Royal Navy's Mediterranean Fleet three days' notice of a large-scale Italian naval deployment. This was the Battle of Cape Matapan in March 1941, during which severe losses were inflicted on the Italian Navy.

*

The success of Allied cryptanalysis of Enigma is well known, but its significance is sometimes overstated. Enigma was not used by the German High Command for its messages, which relied on an enciphered teleprinter system — Lorenz — which we discuss on page 65 and, anyway, intelligence doesn't win battles or wars: it supports the efforts of the members of the armed forces who are deployed to fight. Nevertheless, it gave Allied commanders the information they needed to deploy their forces to best effect and contributed particularly to victory in North Africa and to the success of D-Day.

There is one more factor in Allied success which is not often commented on: the German persistence in using so vulnerable a system. Introduction of the fourth rotor to the Enigma used by the German Navy in the North Atlantic stymied Bletchley Park for a year; a rewireable reflector could have produced significant problems if used more widely; other significant developments were planned. But the Nazi regime never realized that the reciprocity of Enigma was its fatal flaw, and the poor ergonomics of its design, which meant that it took three people to send or receive a message, was never seriously addressed. Germany's reliance on flawed 1920s technology was, for hard-pressed Allied cryptanalysts, a piece of good fortune.

89. ENIG

Four clues are normal. In all remaining clues, the wordplay leads to the answer plus an extra letter that is not entered into the grid. The four answers to normal clues must be encrypted in the order they are given using an ENIG machine to find their entered forms. The extra letters from the other clues spell out a message, but this has likewise been encrypted, and must be decrypted, using the ENIG machine with the settings it has immediately after encrypting the fourth of the normal answers.

Reading the decrypted form of the message will give an instruction telling solvers what to do next.

An ENIG machine works like an ENIGMA machine but has only two rotors instead of three, in this case HCLIONZQVPWJFDXTYUKARGMBES and JZFQBUPTERYMKIANDOGCWHSVLX. There is no plugboard, and a simple reflector carries A to Z, B to Y, etc. This compact notation for the rotors means, for instance, that in the first position of the first rotor there is a wire '7 letters' in length, i.e. long enough to carry A to H. When the rotor is advanced, this same wire will naturally be in a new position, and will therefore be positioned to carry Z to G.

Every input letter passes through both rotors (in an order to be determined), through the reflector, and then back through both rotors in the reverse order. With the rotors in their given initial positions, an input A will arrive at the first position of the first rotor, map to H and thus arrive at the eighth position of the second rotor, where it maps to T, which reflects to G. The letter G is the *output* of the S wire on rotor 2, and S is the output of Z on rotor 1. Thus Z is the encryption of A, with the rotors in this setting.

The right-hand rotor advances by one position after *each* input letter – solvers will need to deduce the initial rotor settings. Whenever the right-hand rotor moves through a specific position (also to be determined), the left-hand rotor advances as well.

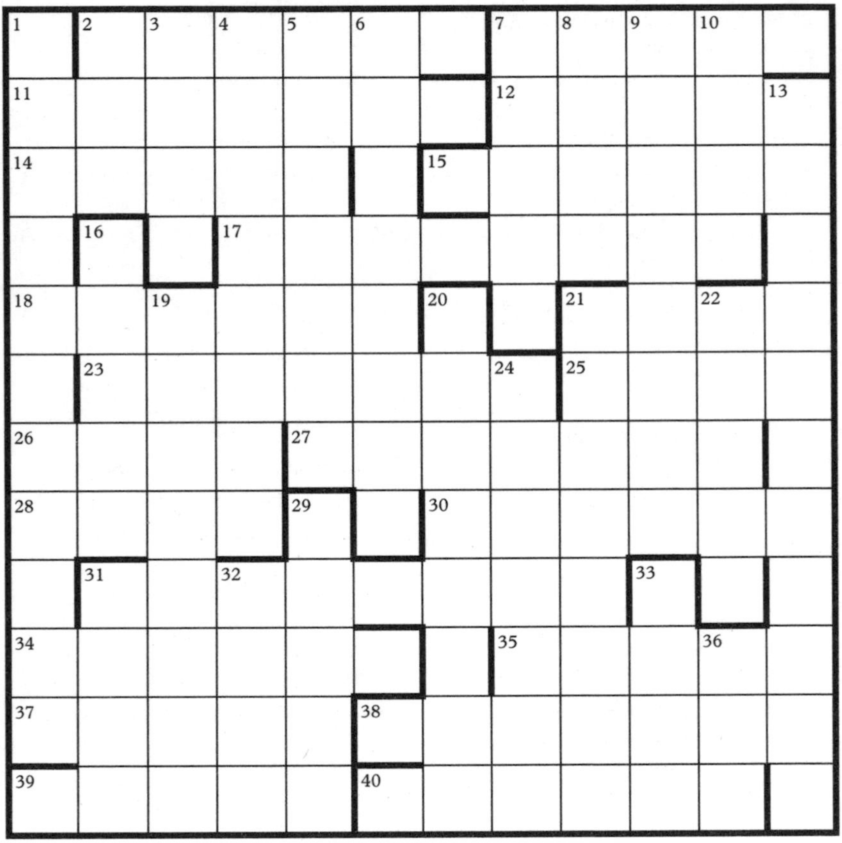

ACROSS

2 Old teacher – success or zilch? Zilch. (6)

7 Dry wind in pampas belt sweeping up centre of Cordoba. (5)

11 Designated route, dry and withered, pursued by famous Okie. (7, *two words*)

12 Measurements applicable to ground where Olympian's included accepted answer. (5)

14 After medium's deduction, unmasks suspect flower girl. (5)

15 Second statement in court is first to satisfy. (6)

17 Ian's screeches, sounds of unhappiness about unfinished artistic endeavour. (8)

18 Deficit keeping America in recession, without secure employment. (6)

21 County in hopeful appeal after run out. (4)

23 Recovers, suffering additional military duties. (7)

25 Old brooch or ornamental choux. (4)

26 Right to enter exposed surface for bit of land. (4)

27 Is NZ honour, in traditional eyes, an excuse for not appearing? (7)

28 Encourages to fight either of two central Asian animals second. (4)

30 Terrible ruler seizing corrupt African city. (6)

31 Flag's displayed by pin, not a sheet of plastic. (8)

34 One mistake after another, that's fate of West centrally. (6)

35 Eastern therapy that's like spiritual healing, possibly ickier. (5)

37 Collection of branches grows out of rubbish, excepting four. (5)

38 Beds location to find crushed beetles without oxygen. (7)

39 Cross I've nearly put into cover. (5)

40 Is abandoning a problem a minor change? (6)

DOWN

1 Joining party that's not left in travel towards Orient. (11)

2 Lease for fixed rent – fine former partner? Everyone should be okay with this. (3)

3 Salmon's surmounting stone to survive. (4)

4 Removals from residence at bottom of dilapidated square. (8)

5 Hollow shell losing height when carried on swell. (7)

6 Apoplectic pilgrims taking in what's kept down in Arles. (8)

7 See about two parts of altered sound. (5)

8 More than one expanse of shifting dunes in fractured range. (4)

9 Predator viciously undoes a second husband. (8)

10 Salts away a pile in different parts. (4)

13 Following African people, English initially singing songs on board. (11, *two words*)

16 Berry about to land beneath ivy branch that's dropped odd bits. (5)

19 Seaside resort finding rovers relaxed, not troubled. (8)

20 Gap in brain membranes accommodating sulci in convolutions. (8)

21 Lead story about unusual term for lesser king and queen. (8)

22 Old coin cast with bits of earlier silver melted down. (5)

24 Less drunk as quarter is awash in lager. (7)

29 Writer's exclamation of greeting – an old motto. (5)

31 Look to paddle initially working up to find water suitable for swimming. (4)

32 Beginner to join Scottish entering North Britain. (4)

33 Travel documentation – woman's heading for South Africa. (4)

36 Greek island Sweden won fairly, but returned. (3)

90. The Ire of Queens

Which two words should be together?

Fin, Found, Green, Ice, Jut, Lap, Mary, New, Rut

91. A complete set

The following set of 6 questions appeared in the 2017 Kristmas Kwiz. They appeared separately throughout the kwiz and there was no hint that they formed a set. The question numbers used below are the question numbers they appeared at in the kwiz:

Q4. This style of question commemorates *Round Britain Quiz* on BBC Radio 4: the format is one long cryptic question which has six parts to it, indicated by letters in parentheses. To gain full marks you should identify all six parts. This will be sufficient to answer the question.

A charity named after a town (a), a Cambridge college (b), and a series of films (c), overlap with a proposition (d), something with horns (e), and either of two complementary time periods (f). How is this?

Q10. The following can be divided into 9 pairs with one left over.

(a) What are the pairs?

(b) What does the one left over represent?

BVCKD, DDCML, DSCRG, FLRNT, FVRTS, KRYTC, LCRTN, MNDCS, NMRTN, NPTCL, NSCBL, PPLRS, PRCRS, PRCTN, PRSSN, RHMTD, TMBRN, TPTNT, XHSTN

Q13. (i) Pair the answers to these clues:

(a) Joint (5)   (n) Tidy (4)

(b) Courageous (5)   (o) Bovines (4)

(c) Calcium carbonate (5)   (p) Irritation (5)

(d) French mustard town (5)   (q) Jest (4)

(e) Engrave (4)   (r) Reimburse (5)

(f) Chilled solid (5)   (s) Smell bad (5)

(g) Formal dress (4)   (t) Educate (5)

(h) Journalist (4)   (u) Employs (4)

(i) Topic (5)   (v) Part of speech (4)

(j) Enlist (4)   (w) Incorrect (5)

(k) Work wool (4)   (x) Chemical element (5)

(l) Minor road (4)   (y) Time period (4)

(m) Intended (5)   (z) Nothing (4)

(ii) What can you make of what's left?

Q15. Answer the clues. All answers have 5 letters. What are the missing words?

(a) Hobbles   (k) –

(b) Equivalent to   (l) All

(c) One of five letters   (m) Desert wanderer

(d) O3   (n) –

(e) Water nymph   (o) Gem

(f) –   (p) Lift using ropes and pulleys

(g) Provide the kit   (q) Divinatory cards

(h) Implicitly understood   (r) Hindu ascetic

(i) Teuthida   (s) Glaze on a cake

(j) Encourages   (t) Baffled; discoloured

Q17. What is the full title arising from:

CIRCUMSTANCE, JULIET, SCRATCHY, FALL, IVORY

Q19. What is the answer to this question? Explain precisely why.

92. Group the groups

Split the following into 6 groups of different sizes (i.e. one group of 1, one group of 2, etc):

ANDREA	ANNE	BERNIE	CAROLINE	COLEEN	ISAAC
JACKIE	JERMAINE	JIM	KAREN	LINDA	MADONNA
MARLON	MAUREEN	MICHAEL	RANDY	RICHARD	SHARON
TAYLOR	TITO	ZAC			

93. Which?

The following list of 55 words can be divided into 10 sets, all of different sizes. Put another way, there is one set of 10, one set of 9, etc. Which word is in the set of one?

A	ABANDON	ABATTOIR	ADAMS	ALDRIN
ANTISEPTIC	APART	ART	AUTUMN	BLIND
CANOPUS	CARD	CHIN	DAYS	EAST
EONS	EROTIC	GAVE	HELIUM	I
INTERSEPTAL	LARGE	LEO	LINUS	LITTLE
MEN	MICE	MONACO	MY	NAP
NAPE	NOTABLE	OLE	PARIS	PENNSYLVANIA
PIGS	POLE	QUEEN	RAPHAEL	REE
REMOVE	RIDGE	SEPTEMBER	SLAIN	TEMPERANCE
THYMINE	TOME	TOWERING	TRANSEPT	TRICK
TRUELOVE	ULSTER	VESTA	WEEK	WISE

94. Christmas songs I

(a, b) JOYEOTHYWAROD

(c, d) FIOTTETREMNEWBAY

(e, f) TOEDOILGAEDCHSIAY

(g, h) DONEDLNTMNRHIMYUNTIIH

(i, j) PSEWSOCMMKHSMIFGRAHTICTAAS

95. What changes?

SPQR historian	_	_	_	_		_	_	★	_	★
Ex big tent baker	★	_	★	_		★	_	_	_	_
Kissed a girl	★	_	_	★		★	_	_	_	_
X Factor Winner	★	★	★	★		_	_	_	_	_
Ex Chelsea defender	_	_	_	_		★	_	_	_	_
Democratic nominee	_	_	_	_		_	_	_	_	_

96. Seconds out

Each part consists of the second letters in the names of the members of various sets. Identify the sets. The individuals may be real or fictitious, and in total there are 8 women, 11 men, 9 animals and 14 other (who are male).

(a) ORT

(b) AAOU

(c) ACEHR

(d) AAAANN

(e) AAOOLNR

(f) AAEIIOOW

(g) AAEEIIORR

97. Odd one out II

What is the odd word out?

Galaxy, Land, Nation, Odyssey, World

98. Sums

(a)

$$\sqrt{\dfrac{\text{nod}+\text{empty-mind}}{\text{digit-shy}}} = ?$$

(b) $$\dfrac{\sqrt{\text{Beverly Hills} - (\text{Blake's} \times \text{Minute Meals})}}{(\text{Heaven Suspect})^{\text{Earth}} - \text{Warehouse}^{\text{Earth}}} = ?$$

(c) $$\dfrac{\text{Men and women every day}}{\text{Caught in the middle of}} \times \dfrac{\text{Years when I wrote this song}}{\text{Paths you can go by}} = ?$$

99. Up and down

Here is a sequence:

1, 31, 311, 2133, 11111, 411114, 2112111, 41141111, 121111131, 8181131112, 511411221, 11411111, 3111111, 111144, 31114, ?, 114, 11, 1

(a) What is the missing member of the sequence?

(b) The last member of the sequence, 1, is really in the wrong place. Where in the sequence should it be?

(c) Explain, precisely, what the sequence is. Be complete in your answer. You may find the following pairs of numbers helpful: 2129291246 and 4122969612, 3694494 and 2949699, 126686 and 212862.

100. Identify the albums

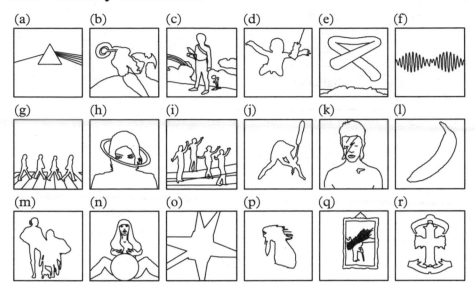

(a) (b) (c) (d) (e) (f)

(g) (h) (i) (j) (k) (l)

(m) (n) (o) (p) (q) (r)

101. P_ _ _ _Y

If H _ _ _ U = 3 and L _ _ _ _ _ _ K = 5,
then S _ _ _ _ T = ?

102. Where?

Where does GROWL fit in the following list? (Read left to right, top
to bottom)

JENNY	MAC	SKIM	LITOVSK	HEAD
TIN	GARDENER	ME	DESPERATE	DENT
SION	MUSTARD	SPIRIT	RUMP	DOC
MINE	CEILING	DETECTIVE	PM	O'FLAHERTY
GILLIE	WIRELESS	ESSE	IRON	PHONE
DOT	CENT	UNSINKABLE	VICTORIA	PESHWARI
REAGAN	BOT	SOLO	WORK	LENINIST
MIST	MUNG	ETHENOL	BUST	SAUCE
HILTON	STRONG	LILLIPUTIAN	FACE	NOSE
GENEVESE	MARXIST	DRAGON	MONKEY	ARC
NINETY	SCHOOLBOY	LAUTREC	GROWTH	LING

103. Scorpion & bee wings

(a) Unadvised rarity

(b) Chewy vendors

(c) Payable school

(d) Risky appeal

(e) Dead copyist

(f) Snogs rabidly

(g) Endowment lever

(h) Took hot glue

(i) Aha, hostesses

(j) Urinal prep

(k) Violent healing

(l) Itemise thongs

104. A gap

 Which country fills the gap?

NOM, SUET, ?, HURTS, FIR, SUTRA, NUS

105. Getting one over

Feedback indicates that our clues can be hard to find. In this case we're intentionally trying to get one over u.

```
09101 12501 42857 16202 ?????
```

COLOSSUS — THE FIRST MODERN COMPUTER?

As technology to encrypt messages advanced, Signals Intelligence had to keep up to stay relevant. And so the move to electromechanical encryption, enabled by machines such as Enigma and Lorenz, ushered in a new era of machines built to mechanize the task of decryption — the world's first examples of electronic, programmable computers.

At Bletchley Park, a young mathematician, Bill Tutte, was assigned to the research section concerned with ciphers that could not yet be exploited and found himself working for the leading cryptanalyst, John Tiltman. The brilliance of Tiltman and Tutte in recovering the design of the Tunny device (as the British called the German Lorenz cipher machine, and its traffic) also showed that it had significant flaws. Bletchley Park jumped upon the task of consistently attacking Tunny messages. Using Turing's ideas, seasoned cryptanalysts under the command of Major Ralph Tester were able to start breaking into messages where some of the machine settings overlapped with those of other messages. Traditional skills of intuition and knowledge of language combined with statistical ideas to resolve the two main challenges: wheel breaking (working out how the wheels were configured from month to month) and wheel setting (working out the start positions of these wheels message by message).

One person relatively new to Bletchley was the mathematician Max Newman. He found that he was not as adept as the members of the 'Testery'. He was frustrated that the task could neither be automated nor approached without the required overlapping. Perhaps inspired by the bombe machines, Newman worked up a proposal using the ideas of Tutte to mechanically test for wheel settings. The proposal was well timed as the Germans were continually becoming more careful in their use of Tunny. Newman proposed complex electromechanical devices that would compare two paper tapes: one with a Tunny message and the other with possible settings of some of the Tunny wheels. The comparison would take place at 2,000 letters per second (a processing speed of 2kHz) and allow wheel setting without overlap. The machines were dubbed Heath Robinsons after the cartoonist famous for his baroque designs.

Tommy Flowers, an engineer from the Post Office research group at Dollis Hill, was brought in to help with the design of the Heath Robinsons. He had already impressed the cryptanalysts at Bletchley with his proposal for bombe improvements. He was unconvinced by the Robinson design, knowing from his own research that purely electronic systems could outperform electromechanical systems in both speed and reliability. On his own initiative he designed a system that would replace tapes with electronic components and that would operate at 5–10kHz.

In a minute dated 1 March 1943 Newman asked for resources for the Robinsons, but also the more speculative design by Flowers. The Robinsons were delivered, but work also progressed on the electronic device that became known as Colossus. Flowers reworked his design to allow a single paper tape, but was still able to achieve speeds of 5kHz. In January 1944 the first Colossus was delivered to Bletchley Park and became part of the 'Newmanry' – which now might be described as an IT department – which grew to 325 employees including cryptographers, operators, engineers and administrators. Flowers's device worked as claimed and the cryptanalysts saw that this powerful machine was capable of much more general computation.

Newman began to envision a complete, fully automated attack on Tunny, and members of the research section developed more complex comparisons that would allow this. Flowers was commissioned to produce more Colossi with more powerful capabilities. The Colossus II was delivered under intense pressure on 1 June 1944 and was able to be configured much more generally than the prototype. The cryptanalysts continually came up with cleverer ways of using the device, including wheel breaking. Subsequent models allowed parallel processing (increasing the speed to 25kHz).

A Bletchley veteran wrote: 'It is regretted that it is not possible to give an adequate idea of the fascination of a Colossus at work: its sheer bulk and apparent complexity; the fantastic speed (of the) thin paper tape round the glistening pulleys; the childish pleasures of ... gadgets; the wizardry of purely mechanical decoding letter by letter (one novice thought she was being hoaxed); the uncanny action of the typewriter ...; the stepping of display; periods of eager expectation culminating in the sudden appearance of the longed-for score; and the strange rhythms characterizing every type of run.'

Demand increased as Tunny traffic became more important, but at the same time the Germans' usage of Tunny improved. By the end of the war ten Colossi were in operation. The research section had shown how Newman's complete attack could be realized on the Colossi. The configurable devices were the world's first examples of electronic, programmable computers; recent work has shown that if the ten Colossi were connected then they would have been able to implement a universal Turing machine, the modern hallmark of general-purpose computing.

After the war Tommy Flowers helped to develop ERNIE, the first machine to generate random winners for Premium Bonds. Max Newman was awarded both the Sylvester and the De Morgan Medals — the two most prestigious mathematical awards in the United Kingdom.

106. Enemy messages

One way to encipher a message is by using a *one-time pad*, in which a random stream of letters (taken from the pad) is used as the key. The sender 'adds' the message to the key, using A=0, B=1, ..., Z=25 to convert letters to numbers and back again (subtracting 26 if the total goes above 25).

For example, if the message is 'HELLO' and the key is 'QWERT', then the cipher message is 'XAPCH':

H=7, Q=16, H+Q=23=X
E=4, W=22, E+W=26=0=A (subtracting 26)
L=11, E=4, L+E=15=P
L=11, R=17, L+R=28=2=C (again, subtracting 26)
O=14, T=19, O+T=33=7=H (again, subtracting 26)

The recipient, who has an identical pad, can then 'subtract' the same key from the cipher to recover the original message. It is not a good idea to use a predictable key (e.g. in the above case we might deduce that the key continues 'YUIOP'), or to use the same key twice for different messages, as this can help adversaries deduce both the message and the key. This kind of mistake led to the first breakthrough into the Tunny cipher – and eventually to the need for Colossus.

We intercepted the following messages – the first from the enemy HQ to some of their army sections, the second a reply from one of those sections – and, judging by the identical starts, we think they may have slipped up and used the same key in both messages! When are they planning to attack?

Message 1:
FCSJH DWFAE DTJMY GGICH RVNOK TUEJO XUGNT RALLU
UYZLY PZCYA VLVV

Message 2:
FCSJS RFKQO OTXCW JRXDO UEYKY IFEYE NMKVV
QAFVO FRPPE IWOU

107. Odd one out III

Which is the odd one out?

(a) ABHORS, ALMOST, BEGIN, CHINTZ, WRONGED

(b) CHIPS, PIE, SODA, SPONGE, TOFFEE

(c) BREAD, CABBAGE, DESSERT, PUMPKIN, WATERCRESS

(d) KIMONO, MUKLUK, MUUMUU, POMPOM, SWEATER

(e) CORMORANT, EXOCET, GADFLY, PHOENIX, VULTURE

108. Which?

The following list of 55 words can be divided into 10 sets, all of different sizes. Put another way, there is one set of 10, one set of 9, etc. What word is in the set of 1?

AGE	ALAN	AM	ANGEL	ARE
BEAST	BISHOP	BRILLIANT	CADET	CORNELIUS
DALLAS	DAVID	DEE	DIVE	ECOLOGY
ENCIPHER	END	EPICALYX	EXPLORATION	FILM
FOG	FORGE	GAMBIT	IMPRISON	INTRODUCTION
JERUSALEM	JUBILEE	LONDON	M	MACARONI
MARROW	MARVELLOUS	MILTON	MIMIC	NIGHT
NUISANCE	OBEY	OWE	PEA	PORT
PURPLE	REHOBOAM	SEA	SHUTTLE	SHY
SPRING	STORM	SUPERB	TIRIEL	TORTOISE
VALA	VULCAN	WONDERFUL	XEHANORT	ZORG

109. Find the word

What word?

MADISON

SATURN

DO

NITROGEN

EXODUS

110. Changing places

If Yemen: Bl → B is the Netherlands, then identify the following:

(a) Scotland: W → Y

(b) Norway: W → Y

(c) Bangladesh: G → B: R → Y

(d) Vietnam: R → B: Y → W

(e) Devon: W → B: Bl → W

(f) Country + Year

(g) Country: R → Y: G → R: W → G

(h) Country R ↔ G

111. All wrong

In this question each digit has been replaced with a different digit, and each operation + − × / has been replaced with a different operation.

45/3 = 64

66×0 = 417

82+9 = 49

49−8 = 6

What is 40/7?

112. Divide into pairs IV

Divide into pairs:

ASIA, BEER, BELL, BREW, CROW, EAR, EASE, HOE, JAW, JEER, MALAISE, MONARCH, MOULD, NIGH, OVER, QUEUE, SIR, VENICE, WEIGHT, WHALER

113. Football logic puzzle III

In the following table every pair of teams has played each other once.
The table is ordered in the usual way (3 for a win, 1 for a draw) with
goal difference, goals for and fair play record used as tiebreakers.
Using arithmetic and logic, fill in all the gaps.

Team	Points	Goals For	Goals Against	Goal Difference
City		5	6	
Rovers		2		
United		1		
Wanderers				0

Rovers	v	Wanderers
City	v	United
United	v	Wanderers
City	v	Rovers
United	v	Rovers
Wanderers	v	City

114. Octets

Put into sets of eight:

(a) AGUKQCTR, AKT, AKTDUA, BAKKAI, BLRRLT,
BMGCTR, CMPQAPVTDQC, COOSL, DYLKA, ECMHKJE,
ESPTMGH, GVEEVLA, IOATE, IRQYG, LHLPDAKT,
LRKQLRQKICJSKJCTR, LVPIDG, MJBYEQCTR,
MOHPDEK, NAT, NYHQCTR, RERJCTR, TVDHMK,
VQNVPVLSQ

(b) BL, CN, DM, FG, FT, GL, GN, GN, HT, IH, KG, LK, ME,
OE, PD, PE, PK, PT, QR, QT, RD, SE, TN, YD

115. A small mix-up

Who are these people?

Jenny Soles, Jilted War Hall, Congenial Pink Hen, Weirder Spread

116. Some explanations

Explain the following:

(a) FDWBKRHQIRELAND

(b) TIVUCPENCE

(c) SIBUHEW

117. Christmas sudoku

One of the rows (reading left to right) is the same as one of the columns (reading top to bottom).

R				H				
						S	S	
					I			
			R					
					C			M
S	I						A	
					M	A		
	H				T	I		
			S			M	R	S

118. Not a TDous question we hope!

If FT=GD and SD=SR, then what is TD?

119. Next in sequence/fill in the gaps

Identify the '?' The '...' represent missing entries that you are not expected to provide. The '–' indicates that there are no possible values for that entry. In (e), the values are exact.

(a) 2, 6, 7, 5, 9, 10, 8, 40, –, –, 12, 20, 14, 18, 16, 70, ?, ?, ?, ...

(b) 100, 200, 300, 301, 302, 303, 304, 309, 350, 351, 352, 353, 354, ?, ?, ..., 938, ?, ...

(c) New York, Hawaii, Connecticut, Arkansas, Massachusetts, Illinois, Georgia, Nebraska, ?, ?, ...

(d) 15145, 202315, 2081855, 6152118, ?, ?, ...

(e) .349, 9.3, 529, 5.433, .974, ?, ?, ...

120. Numbers and letters

(a) Identify the following:

26830 PG of TTT

5 T on a F

3 F in a Y

1 YOA is about 2.5 P

(b) Identify the following where the letters in bold have been encrypted:

330 **R** of **O**

370 **R** of **RQ**

415 **R** of **SQ**

440 **R** of **E**

494 **R** of **M**

554 **R** of **AQ**

622 **R** of **JQ**

(c) Identify the following where the digits have been encrypted:

7, 51,   ABFH

5, 4,   BMS

55, 54, DAD

2, 0,   KATD

6, 9,   LTS

52, 50, MA-C

53, 58, MITK

56, 59, MIW

57, 41, MPIE

3, 8,   PUS

(d) Finally, we have encrypted both the letters and the digits:

The	**S**	has	**8 E,**	**0 H**	and	**8 U**
The	**L**	has	**0 E,**	**65 H**	and	**7 U**
The	**J**	has	**7 E,**	**65 H**	and	**0 U**
The	**Y**	has	**65 E, 21 H**		and	**51 U**

What does the **A** have?

121.   What's next?

What comes next:

M, N; A, I, W; C, O; ?

122.   What's happened?

What's happened here?

CHKLET, KRIMKCK, TOLSARIA, TKTERMKRY,
WELLIMSTKM

THE ARRIVAL OF THE AMERICANS

Today's threats are global and communications are complex;
but Signals Intelligence has always benefited from a
collaborative approach between allied countries. The Sigint
partnership between the UK and the US is so much a part of
'business as usual' that we rarely pause to reflect on the
fact that it has a precise and defined starting point.

After the fall of France in the Second World War, President
Roosevelt sent William Donovan, the future leader of the
OSS (Office of Strategic Services), to the UK in July 1940
to assess whether the UK was willing and able to continue
the war against the Axis powers. He returned to the United
States convinced that Britain could continue the fight,
but only with support from the US. The immediate result of
his return was the signing of the agreement by which fifty
US Navy destroyers were transferred to the Royal Navy, in
exchange for the UK offering to the US ninety-nine-year
leases on naval and air bases in British territories in the
Americas. At the same time a small group led by Sir Henry
Tizard travelled to the US to discuss the possibility of
using America's industrial might to mass-produce highly
advanced technologies such as radar. It was in the context
of these talks that the US surfaced a proposal to exchange

cryptanalytic information about the Germans, Italians and
Japanese.

The proposal was met with lukewarm interest in the UK.
The first Turing—Welchman bombes were in operation,
and Enigma messages were being decrypted. The secret
of the vulnerability of Enigma was closely protected,
and nobody could see any advantage in giving this to the
Americans. The Americans offered to send a team to the
UK to discuss cryptanalysis and A. G. Denniston, head
of GC&CS at Bletchley Park, agreed, reluctantly, to hold
talks of a general nature with them. The Prime Minister,
Winston Churchill, gave his consent but expressly forbade
discussion of Enigma.

The US side had, unbeknown to the UK, their own 'big secret'
to share. They had been successful in breaking the Japanese
diplomatic cipher system codenamed (by the Americans)
Purple, and they would bring with them a machine they had
devised ('The Purple Analog') to decrypt Purple messages.
It is likely that very senior Americans might have thought
that this went some way to match in value and importance
some of the advanced items the Tizard Committee had shared
with the US side.

The US Navy and Army each chose two men to send to the
UK: Abraham Sinkov and Leo Rosen for the Army, Robert
Weeks and Prescott Currier for the Navy. They would bring
a Purple Analog with them as a gift to the UK. Their
journey was, to say the least, eventful. They embarked on
the battleship HMS *King George V*, which was returning
to the UK after having dropped off Lord Halifax, the new
UK Ambassador to the US, on 25 January in Annapolis. At

first they made slow progress as *King George V* was acting
as escort for a convoy, but when the convoy was met by a
destroyer force on 3 February, they were able to make good
speed for Scapa Flow, the Royal Navy's base in the Orkneys.

The group was supposed to fly south in Short seaplanes,
but the crates containing the Purple Analog and other
equipment they would be leaving with the British would not
fit through the doors of the aircraft. They were instead
found room on HMS *Neptune*, which was heading through the
North Sea and the English Channel to Plymouth for repairs
and which could drop them off at the Thames. The equipment
was stored on deck while the Americans stayed warm and
dry below. On their journey south, *Neptune* was strafed by
German fighter aircraft, but luckily there was no damage to
the cargo.

They arrived in Sheerness on the evening of 8 February
and the men and equipment were unloaded. They drove to
Bletchley through London, where they saw the devastation
that had already been caused by the Blitz. Tired,
disorientated and sobered by their first experience of being
in a country that was at war, they arrived late at night at
a blacked-out Bletchley Park, where they were welcomed by
Denniston, Travis (Denniston's deputy) and Tiltman (the
leading cryptanalyst), and a tray containing large glasses
of sherry. They then left for Shenley Park, Lord Cadman's
country house, which would be their base during their stay
in the UK.

During the next couple of weeks the Americans were exposed
to all the work of Bletchley Park except Enigma. In
particular, they saw the way in which the UK was working on

the mechanization of the cryptanalytic process. Tiltman,
who acted unofficially as host to the Americans, saw that
not talking to the Americans about Enigma after they
had presented the UK with a Purple Analog had created an
imbalance, and at his insistence, a request was made to
Churchill to share UK progress on Enigma with the US team.
On 27 February the prime minister gave his permission.

This was the first step in what became an ever-closer
partnership, and which led eventually to more than a
hundred American personnel being stationed at Bletchley
Park. In turn, the relationship that developed at
Bletchley Park was of such value to the UK and the US that
after the war the new prime minister, Clement Attlee, and
the new president, Harry Truman, agreed to its continuing.
The UKUSA Agreement signed as a result in 1946 still
underpins collaboration between the UK, the US, Australia,
Canada and New Zealand today.

123. Letters from America

What word (10 letters, hyphenated) can be deduced from this article about America:

The 'special relationship' reaches back to the early days of the American republic. By the nineteenth century the transatlantic exchange of manufactured goods for raw materials was a cornerstone of the world economy. One cause of the US Civil War was the notion that an independent Confederacy could survive by basing its entire economy on the exports of 'King Cotton'.

The United States has always been a nation of immigrants. The country's sheer vastness helped create a diverse 'nation of regions', reflecting its many native and colonial influences. Along El Camino Real in California you will find a string of Spanish names: Chula Vista (beautiful view), Monterey (king's mountain), Palo Alto (tall stick), Alcatraz (pelican).

We think of America as a young country, but as one of the world's oldest democracies it has long been influential politically, economically and culturally. The Massachusetts Bay Colony was a successful economy by the mid-seventeenth century. In 1652, to achieve more independence from 'mother England', the Boston authorities allowed two settlers to mint coins depicting a key local export: a pine tree.

The rapidly growing republic was a world leader in emerging communications technologies: Thomas Edison's phonograph, Samuel Morse's telegraph, Alexander Graham Bell's telephone and George Eastman's photographic film revolutionized the nineteenth century. In the 1860s transcontinental telegraph and railroad links connected the eastern seaboard to the Pacific Ocean.

In 1941 Roosevelt and Churchill issued the Atlantic Charter, which laid the foundations of the postwar world: the United Nations, GATT, decolonization, and a secret codicil agreeing cooperation on Signals Intelligence. We have stood by each other ever since, through the Second World War, the Cold War, and into the modern era with its global challenges. Down the years, the UKUSA agreement proves that friendship truly is golden.

124. The Americans are coming

The Americans are coming. When they arrive, what will this sequence become?

AGE, STAIN, JOT, TOR, CAMP

125. Find an example

Find an example of a word that could precede the sequence:

```
???????
CRLIAIC
HAFKPMH
OYAEAAO
```

126. Hidden

Find the hidden letters in or implied by each of the following sequences:

(a) ANTIPASTI, ABERNETHY, ARCLENGTH, ACIDIFIED

(b) LEAFLADEN, ABSEILING, TRILLIONS, UNAVOIDED, REFRACTED, MILKSHAKE, GILRAVAGE

(c) CHARLI, BATTLE, (le dernier) LOUIS

And likewise for the hidden number in the following sequence:

(d) 9.07184356, 3.56711228, 4.43322112, 1.65454321, 5.31974286, 4.44432222

127. Numerical order I

Put each of the sets of 5 words or names into numerical order:

(a) Balfour, Fivepence, Telephone, Threesome, Trustworthy

(b) Andrei, Funfair, Olivier, Rosenzweig, Veins

(c) Michigan, Oregon, Phoenix, Santa Ana, Washington

128. A to Z

A = GAG, R, TITA	N = BWEM, IU, N
B = ADB, SAD, WWB	O = BHN, DBTT, DOYS
C = AHFOD, AROBTTH, MX	P = F&B, LWTCDI, NT
D = QAWNMAWAD, S, TD	Q = ANATO, HS, MIH
E = HC, OOTN, LROOE	R = AE, RTR, TTTD
F = AP, CSD, II	S = CATR, GUAW, OTR
G = ATOTT, SEBTP, TLLDOB	T = BTB, PUTJ, R
H = BA, D&B, DA	U = Q, SOR, V
I = K, LLT, SS	V = BB, DTD, HAK
J = E, F, I	W = F, MFTEOH, MIB
K = POKR, SFA, TPI	X = DAW, O&L, S
L = S, S, S	Y = CTTE, GFTO, KTA
M = AOS, BM, CYC	Z = ABOD, GOS, SC

129. Hidden phrases

Each of the following sequences contains a hidden, apposite phrase.
All you have to do is find them.

(a) T3215, A34, U52, J564, C135, A345, Q45, D4

(b) N3, P6, H7, K1, F8564, I34, C1, Y3, K4, C4, F28

(c) V473, F6243, A46, G452, P8, V1, L4, T24, K1, L1, D23, B24,
     E41, A45746, N21, K156, S31

(d) TSAVO, BEBEK, BERNE, ONECO, DIJON, INOLA, ATHUR,
     THAME, CERNA, NEUSS

(e)   ASIDE, IVORY, EPOCH, OFTEN, TORCH, LILAC, TIMID,
       VODKA, SATIN, CHARM, DWELL, JAUNT, EYRIE, USURY

(f)   4th, 6th, 73rd, 61st, 55th, 22nd, 24th, 16th, 10th, 56th, 60th, 63rd, 7th

(g)   4, 3, 1, 7, 8, 4, 3, 5, 9, 10, 5, 6, 10, 3, 2, 4, 8, 5, 3, 4, 4, 9, 2

(h)   31, 63, 41, 59, 54, 19, 77, 51, 36, 53, 49, 11, 73, 67, 37, 33, 72,
       21, 62, 20, 19, 7, 16, 29, 75

130.   Musical finale

What completes:

Emmanuel, Atkins, Dylan, Gilmour, Beck, ?

131.   Street snooker

In a recent game of street snooker, one of the setters achieved a
maximum break of 125.

The game began ...

SD, F, SD, F, SD, F, SD, F, SD, F, SD, F, SD, F, SD, F,
SD, F, SD, F, SD, F, SD, F, SD, F, SD, F, CS, GT, ...

... when right on cue we knocked the kwiz koffee kup over on to the
score card, leaving it unreadable. What are the final missing entries?

132. Categorical connections

(a) Planet of the solar system, _____ , Bananarama single

(b) Humphrey Bogart movie, _____ , Moroccan city

(c) Brown, _____ , _____ , Space shuttle

(d) Madonna no. 1 single, _____ , _____ , 101 Dalmatians

(e) Genesis album, _____ , _____ , _____ , Fizzy drink

(f) Sliver, _____ , _____ , _____ , Silver

(g) Joshua Reynolds's painting, _____ , _____ , _____ ,
US state capital

133. Divide into pairs V

Divide into pairs:

(a) 1, 2, 5, 8, 10, 12, 24, 28, 32, 36, 40, 56, 81, 84, 92, 96, 163, 204,
655, 768

(b) 1, 4, 9, 21, 22, 33, 37, 202, 205, 241, 333, 462, 717, 962, 1606,
4649, 13837, 333667

and what is the answer to this one?

(c) 3, 7, 10, 14, 32, 100, 142, 294

134. Crossword

Across

3 Cross _____ or _____
   Down (8)

5 Across (12)

7 Down (10)

8 Across (8)

11 Across (4)

12 Down (6)

13 Worn down (5)

Down

1 _____ Down (2015 single
   by Romanian singer) (5)

2 _____ Down (sequel to
   1977 Cold War novel) (7)

3 Down (10)

4 Down (5)

6 Down (3,2)

9 Down (3,2,5)

10 Across (6)

135. Kwiz function

In mathematical notation fg(n) means that we apply a function g to a number n, and then apply a function f to that number. If:

ik(1)=2

iw(9)=9

iz(11)=4

ki(13)=5

kw(15)=53

kz(16)=257

wi(18)=9

wk(19)=32

wz(20)=4

zi(22)=16

zk(24)=841

zw(26)=3844

then what is kwiz(23)?

136. Who?

Who (currently) completes this sequence?

WH, PT, JP, TB, PD, CB, SM, PM, CE, DT, MS, PC, ?

137. FAQ

Who comes next?

WW, AR, DR, RG, BR, GT, TV, GH, KK, SE, SM, FC, RH, SA, ?

138. Christmas songs II

A, A, A, ABOVE, AFRAID, AN, ARE, ARE, AT, A-WASSAILING,
BE, BE, BELLS, BELLS, BELLS, BRIGHTLY, CAROLLERS,
CHESTNUTS, CHILD, CHRISTMAS, CHRISTMAS,
CHRISTMAS, CHRISTMAS, CHRISTMAS, CHRISTMAS,
CHRISTMAS, COME, COME, COME, DASHING, DAY,
DREAMING, EARTH, FIRE, FIRST, FROM, GOOD, HARK,
HAVE, HEAR, HEAVEN, HERE, HOLY, HOW, I, I'M, IS, IS,
IT'S, JINGLING, JUST, KING, KING, LISTENING, LITTLE,
LOOKED, ME, MERRY, MY, NEED, NIGHT, NO, O, OF, OF, ON,
ON, OPEN, OUT, PA, PUM, PUM, PUM, RING, ROASTING,
RUM, SAID, SHINING, SING, SLEIGH, SLEIGH, SNOW,
SNOW, SO, SONG, STARS, THE, THE, THE, THE, THE, THE,
THERE'LL, THERE'S, THEY, THEY, THIS, THOSE, THROUGH,
TIME, TO, TO, TO, TOLD, WE, WELCOME, WENCESLAS,
WHITE, YOU, YOURSELF

139. Enigma Variations II

Fill in the gaps such that once complete each line will in turn lead
somehow to the letters E, N, I, G, M, A.

(a)   The nymph was in love with _ _ _ _ _ _ _ _ _ _ .

(b)   The poem was written by _ _ _ _ _ _   _ _ _ _ .

(c)   The country is one of the _ _ _ _ economies.

(d)   The car is manufactured by _ _ _ _ _ _ _ _ _ _ .

(e)   The schoolboy novel is by _ . _ . _ _ _ _ _ _ _ _ _ _ .

(f)   The particle is a _ _ _ _ _ _   _ _ _ _ _ _ _ .

140. As easy as un, dau, tri

Where is this?

(1) Cardiff

(2) Neath

(3) Swansea

(4) Holyhead

(5) Hay

(6) Bridgend

(7) Newport

(8) Montgomery

POST-WAR GCHQ –
WHY CHELTENHAM?

Today GCHQ is located in Greater Manchester, London,
Bude and Scarborough but it is most famously associated
with Cheltenham, where its main base is. GCHQ has been
in Cheltenham for so long that few people realize that it
wasn't inevitable that it would end up there.

In the months following the end of the Second World War
there was a massive reduction of staff at Bletchley Park:
from about 10,000 at the end of 1944 to fewer than 2,000
at the end of December 1945. The site at Bletchley was far
too big for post-war GCHQ (the new name for GC&CS), and
much of the space was needed for the Post Office, so GCHQ
had to move.

The expectation was that GCHQ would return to central
London. Until August 1939 the organization had shared
premises with the Secret Intelligence Service (MI6) by St
James's Park at 54 Broadway but the post-war organization
would not fit into any available accommodation in central
London. GCHQ therefore moved on 1 April 1946 to Eastcote,
a former bombe outstation in the suburbs. Eastcote was
a stopgap, and was not ideal: it wasn't big enough and so

several of the London premises occupied during the war also had to be used; it was a long way from central London, in many ways as inconvenient as Bletchley had been; there were few amenities locally; and some of the staff who had been prepared to move from Broadway to Bletchley Park during the war now wanted to return to central London.

By mid-1947 a rethink in government policy meant that GCHQ needed to move away from London. A lesson identified from the V-2 Blitz of 1944–5 was that the concentration in Whitehall and its immediate neighbourhood of government departments that would be crucial in wartime was a major vulnerability, and that in the face of a threat from ballistic missiles these departments should be dispersed. GCHQ, therefore, was asked to think about where it might go. Its principal need was good communications: the war had shown that electronic collocation between the Sigint Centre, its outstations, the service ministries and deployed Commands worked well, but the telecommunications requirements were high to begin with, and would grow. Furthermore, there was likely to be some competition with other government departments for what capacity existed in London.

A planning document of September 1947 listed the considerations that needed to be taken into account: the southern half of England was preferable, though built-up areas and the coast should be avoided; good communications (in all senses of the word) were essential; and reasonable accommodation for a large number of staff would be necessary.

*

Both a return to Bletchley and a move to Cambridge were discounted because of the competition for telegraphic cables. Taunton and Shrewsbury were discounted because of the sparse cable infrastructure. Somewhere in the Manchester—Liverpool area might have been promising but the planned expansion of industry would mean competition for cables. The most favourable areas looked to be Gloucestershire and Wiltshire: an extensive network of landlines had been set up during the war to meet both British and American requirements, and these were under-used at the time the report was being compiled.

In October 1947 a member of GCHQ, while on a private visit to Cheltenham, heard that there was a large set of government buildings at nearby Benhall Farm, currently occupied by the Ministry of Pensions but likely to be vacated within a year or two because the majority of the ministry's staff and operations were in Blackpool. He arranged to look over the site, presenting himself as an Admiralty official interested in pensions procedures.

His report was most encouraging. Capacity was said to be sufficient for 3,500 people; the six blocks appeared to be sound; decoration was good; a canteen catered for the present on-site staff of 2,000; the site had a number of supporting buildings for garages, and a Ministry of Works depot adjacent; there were buses to and from Cheltenham every ten minutes; and the local council, businesses and people were already well used to civil servants.

There was a second site on the other side of Cheltenham at Oakley Farm, in temporary use as a teacher-training college. The two sites had been the Logistics Headquarters for the US Army and the extensive network of landlines built for them was another major attraction of Cheltenham. From this moment onwards, Cheltenham was the only location seriously considered by GCHQ.

141. Sextuplets?

45° 54' 58" N 6° 07' 59" E

40° 04' 00" N 75° 06' 59" W

51° 32' 02" N 9° 56' 08" E

43° 35' 07" N 39° 43' 13" E

37° 30' 47" N 122° 07' 13" E

142. Order then reorder

Put the following in the correct order from slowest to fastest:

AN INN TOAD
GOTH ALERT
ITS PROMISES
MAIN OAT
MORE TOAD
POSTER

143. Properties I

Each of the items to the left of the colon has a property that the item to the right of the colon doesn't have. What is the property?

(a) 2, 3, 23, 109, 139, 1103, 1303, 1327, 11777, 13873 : 1277

(b) (11,23), (13,19), (17,83), (19,79), (19,97), (23,47), (29,71), (37,67), (37,79), (43,97) : (47,71)

(c) (1,8,10,39), (2,6,73,75), (3,22,31,52), (4,5,8,15), (5,39,57,74), (6,33,53,102), (7,8,34,39), (8,16,53,68), (9,53,60,68), (10,16,81,99) : (11,53,58,111)

144. Genned up

If Bar=02, Sap=10, Reg=14, Far=18 and Rag=78, what does Gen equal?

145. Take a chance!

Which two items are missing from this list?

–, BER, BZL, CHN, EEN, EST, FGHNTN, GAT BITAIN, GLD, GNN, ICLAND, JPN, KAIN, KMCHTK, L, LB, LSK, M, MDDLE ET, MDGS, MONGOL, ND, NDONE, NEW GNE, NOTH, NTHN, ONGO, P, QUB, RKUTK, S UD SS, SOUTH, STHN, URL, VNZL, WEEN, WS UD SS, WS Y, WSTN, YKUTK

146. Odd one out IV

What's the odd one out?

GARNET, IRON, RIGS, ROGUE, SORE, URBAN

147. Changing situations

(a) AYBS → G&F

(b) TLL → WHTTLL?

(c) C → F

(d) F → J

(e) RAC ← OFAH → TGGG

(f) FF → FF

(g) SOO(P2): PAE → P → GS

148. Who made it?

If INTEND made the I and ON made the IT, who made the BO?

149. **Christmas songs III**

   (a) DCEDFBEBBBBBBBBB

   (b) BDCAEIBDCAEIBDCAEICAECD

   (c) ACBDHBCCADDFBCBDCFCDEAECD

   (d) EDDAEECDCBDBEDDEDCEDCGG

   (e) CCDIDBCDDCDFGCHGCBBDE

   (f) DECEBCGBEBECECDDCCDBCDDBC

   (g) ABCCEGFCIDABCCIDCCF

   (h) CFECCFCCFCDBGCCEEBFBD

   (i) FFCIDBCIECIDECCCEFEBD

   (j) DBEFDEAEFDEAFDCDBAFCCC

150. **Spot the order**

If you remember enjoying Robert Carlyle as Hamish Macbeth, you'll love seeing Jim Broadbent rise from his sickbed to play – which Shakespeare character?

151. **Some numbers**

   (a) Pair:

       ATI, DBJ, EIGHT, EVS, FIVE, FOUR, HDT, LFD, NINE, ONE, ONE, PHN, SEVEN, SIX, SPO, TEN, THREE, TJX, TWO, WRY

   (b) In the list below, all members are equal. Explain.

       BFPFV, DLNNN, FPX, HLLX, JLVD, OAUXH, TXTV, VLR, VXP, XJR

152. Outlines

Each image is the outline of two countries that share a border.
However, the border has been removed. Identify the countries. Note
that not all are presented in their usual orientation. Scale has been
ignored.

(a)

(b)

(c)

(d)

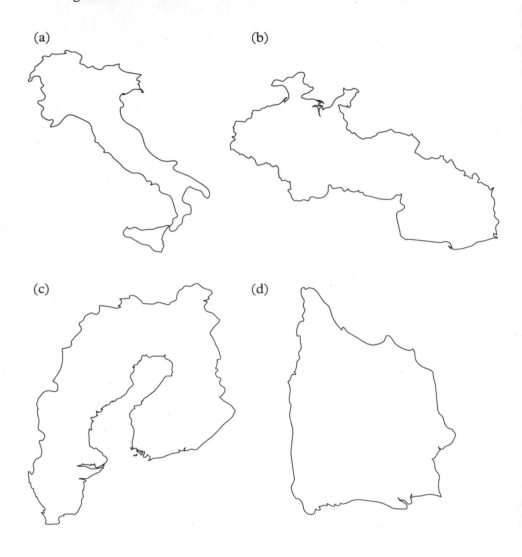

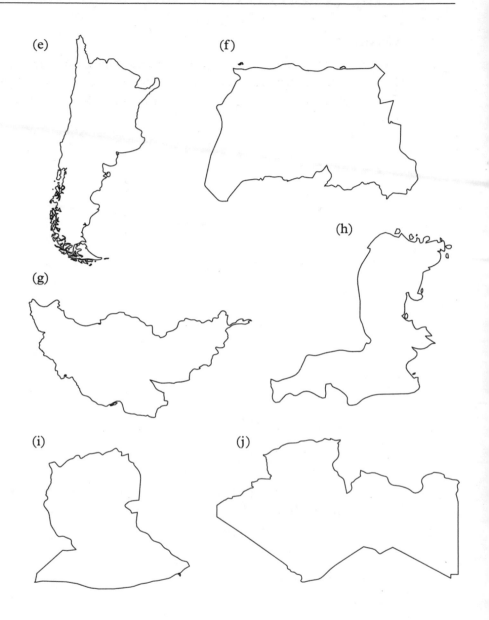

(e)

(f)

(h)

(g)

(i)

(j)

153. Take a break

Explain the order:

HEAD, BELLY, FINGERS, NOSE, BEARD, EYE, LEG

154. Miscellaneous

The parts of this question really are miscellaneous – though are linked by being Trivial Pursuit categories!

Geography
(a) As the crow flies, I am 42.36km from Gloucester Shire Hall, 44.82km from Worcester Shire Hall and 38.26km from Oxford County Hall. What am I and what is the missing distance?

Music
(b) Using the Parsons Code, identify the group who sound a bit like this:

> National Anthem of Paraguay (9)
>
> Schumann: Mit Myrten und Rosen (8)
>
> J. S. Bach: Magnificat in D No. 7 (9)
>
> Schubert, Franz: Symphony No. 9 in C 'Great', 1st movement, 3rd theme (9)

Literature
(c) Identify and divide into pairs: FP, KL, R, RAJ, STM, TN, TT, TTOTS, WSS, 1TIHAY

Art
(d) I am an artist. When I sign my name using predictive texting, the sequence of keys for my first name is the same as the sequence for my surname. Who am I?

155. Sums

(a) (Little Pigs × Billy Goats Gruff × Blind Mice) − Bags Full = ?

(b) $$\frac{(\text{Kamui} \times \text{Felipe}) + (\text{Kimi} \times \text{Esteban}) + \text{Jolyon} - \text{Jules}}{\text{Rio} - \text{Lance}} = ?$$

(c) $$\frac{(\text{C minor} - \text{E}\flat \text{ major}) \times \text{A major}}{\text{B}\flat \text{ major} - \text{D major}} + \text{C major} = ?$$

(d) $£\&-\sqrt{\&^\wedge+(\$\%+\$")/£+\%!-\%^\wedge} = ?$

156. Wordbox (Letterbox)

```
R    E    E    B
O    T    A    C
F    I    P    P
N    E    M    E
K    W    O    R
```

What 4-letter word starting with S is missing from this Wordbox?
It might represent your response to this kwestion!

157. On tour

The setters had a tour of England, to four places not usually visited by tourists. The organizer sent the following messages to tell the others the places to visit. Which places were visited?

E, SW, –, W, S, E, NE, S, N

N, SW, NE, –, S, NW, W, S, NE

S, E, NW, SE, NW, NE, SW, NE, S, –

W, N, –, N, S, NW, SE, NW, SE, SW, –

158. Four to fill

Fill in the gaps:

_ _ _ L I _

_ _ _ L A

_ _ _ L O

_ _ _ _ L E   _ _ _ _

159. Lend me your ears and I'll sing you a song ...

(a) If YESTERDAY is 1ˢᵗ, PENNY LANE is 2ⁿᵈ, A HARD DAY'S NIGHT is 3ʳᵈ, NOWHERE MAN is 4ᵗʰ and SUN KING is 5ᵗʰ, what is OCTOPUS'S GARDEN?

(b) If NORWEGIAN WOOD = 1, YELLOW SUBMARINE = 2, REVOLUTION = 3, BACK IN THE USSR = 4, DRIVE MY CAR = 5, what is THE BALLAD OF JOHN AND YOKO?

160. Sequence II

Fill in the gaps:

1, 2, 2, 10, 2, 10, 1, 2, 2, 10, 2, 10, ?, ?, ?, ?, 1, 2, 2, 10, 2, 10

161. Verbs

Find and answer the question

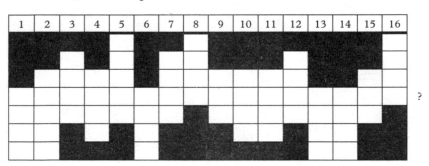

?

1, 3	Relax by provoking (4)
2, 5	Dismiss idea when injecting drugs (5)
4, 8	Support by giving way (4)
6, 16	Make fun of someone to dismiss from university (4)
7, 12	Appear unexpectedly and decline (4)
9, 15	Prepare to put people off bus (3)
10, 13	Need attention to return from high (4)
11, 14	Accompany meal with drink, then clean the plates (4)

NON-SECRET ENCRYPTION –
CHANGING THE RULES

When it comes to tackling problems that on the face of it seem impossible, it takes the ingenuity and teamwork of people who think differently to have success. Alan Turing is often held up as the talisman who embodies this ethos of GCHQ, but there are many others.

James Ellis was a classic example of a divergent thinker at GCHQ. An electrical engineer drafted in from Dollis Hill, Ellis was a voracious reader with a love of cross-disciplinary thinking and a reputation for radical ideas. In the 1960s secure communications devices were proliferating. The challenge arose of how to distribute secret keys safely among them. CESG (GCHQ's secure communications design group) turned to Ellis for new ideas. Ellis instead chose to challenge the premise; did someone need to possess a secret key in order to encrypt a message? In 1969, taking inspiration from a throwaway remark in a Bell Labs technical report, Ellis envisioned a system where the receiver provided all secret confidentiality to a message. Although Ellis lacked the deep cryptographic expertise to design such a system, he could articulate his requirements and knew enough information theory to show that such a scheme was theoretically possible. He wrote down his ideas and looked to cryptographers to help him.

Divergent thinkers can sometimes have more trouble than most communicating their ideas. GCHQ's mathematicians struggled to follow Ellis's talks and papers. Some thought that an argument could be found that non-secret encryption was impossible. The Chief Mathematician, Shaun Wylie, was called in. 'Unfortunately, I cannot see anything wrong with this,' was his guarded judgement. Designers tried to come up with solutions based on common cryptographic ideas, but none succeeded. The idea began to languish as a curiosity of information theory. Again, a fresh idea was needed.

Cliff Cocks had become frustrated with academic mathematics. His doctoral training in modular forms required too much background material to be learned before one could solve problems, and solving problems was what Cocks enjoyed best. Knowing friends who had joined GCHQ, he decided to drop out of his doctorate and try to join them in 1973. A mathematics test and an interview with Wylie confirmed his suitability and he was soon alongside his friends. Within a few weeks of joining, Cocks's mentor Nick Patterson raised the topic of non-secret encryption over tea. Patterson posed the challenge in a very mathematical way: in terms of one-way functions rather than the pragmatism of cryptography. To Cocks it seemed like a mathematical puzzle of the sort at which he excelled. That night in his digs, he set about solving it (purely in his head, as secrecy prohibited writing the ideas down outside of the office). A few months earlier his university supervisor had given a seminar about new ideas in factoring large numbers into primes, a classic example of a one-way function being hard to reverse. A number that was hard to factor seemed a natural starting point for Cocks. Given such a number, Cocks reckoned that there was a limited number of things one could think of when trying and one of his first ideas quickly turned into a solution. The next morning, he wrote up his solution at work and handed it to Patterson.

Patterson grasped the idea swiftly and understood its strength, and he rushed around championing it. Cocks, meanwhile, was content with having devised an elegant solution to an abstract problem that had eluded others. He decided to share it with his old schoolfriend and housemate Malcolm Williamson, who had been one of the people who drew Cocks to GCHQ. Whether out of their good-natured rivalry or the common scepticism about the idea of non-secret encryption, Williamson thought that there was sure to be a flaw. Overnight, he cogitated and came up with a different proposal along more traditional cryptographic lines, which he felt would be as secure as Cocks's idea. Williamson took longer to write up his solution (probably because he was looking for flaws), but both ideas were soon being shared discreetly around GCHQ. The pair of ideas was deemed important enough to be shared among international partners and trusted academics to gain the best possible assurance of their security.

The ideas were also passed to engineers to test feasibility. Cocks and Williamson met with a group of around ten implementation experts at CESG in Eastcote. There were problems. CESG wanted the design hard-wired in circuitry because software was not trusted to behave securely; this

was not well suited to Cocks's idea. More importantly,
designers realized that the sender must be assured that
they are communicating with the intended recipient rather
than an adversary operating a relay: the so-called man-
in-the-middle attack. CESG had no means of solving this
authentication problem and so non-secret encryption was
not used to solve the key distribution problem.

Matters took an interesting turn within a few years.
The ideas behind non-secret encryption began to emerge
in the academic community under the name of public key
cryptography. Williamson's idea, now known as the Diffie–
Hellman key exchange, was published in 1976, and Cocks's
scheme, now known as RSA, was rediscovered in 1977.
Critically, the academic researchers also came up with the
idea of a digital signature that solves the authentication
problem that had brought non-secret encryption grinding
to a halt. There was still not much use for the ideas in
the 1970s, but they did bring about a boom in academic
cryptography. By the late 1980s public key cryptography was
being used to develop secure products such as the US STU-III
secure telephone or the BRENT telephone designed by CESG.
By the 1990s people were putting forward proposals for
secure internet connections using public key cryptography.
Today it is the key enabler for secure transactions on
the internet, protecting billions of pounds of financial
transactions every day.

With the advance of ever-greater computing power, a process
to find new public key methods for the internet is currently
taking place.

162. Factorization

Cocks's non-secret encryption scheme relied on the difficulty of factoring large numbers. The following 100-digit number is the product of two primes about the same size. Can you factor it?

3232320832001656244872162408720872162424644 8163256
444444114400227733669922331199119922333388 66224477

163. Rising sets

(a) If C=4, E=5, T=6, E=7, C=8, M=9, P=10, T=11, then what does U= ?

(b) If J=4, R=5, G=6, then what does P= ?

(c) If L=4, P=5, E=6, then what does S= ?

(d) If HA=10, JK=11, HS=12, YS=13, then what does DC=NC= ?

164. Changing capitals

From CAIRO to PARIS is distance 4; from PARIS to MADRID is distance 3; from MADRID to BERLIN is distance 5; from BERLIN to DUBLIN is distance 3. What is the closest capital to LONDON and how far is it?

165. Divide into pairs VI

Divide into pairs:

BEGINS, BRANCH, DELIUS, GAWAIN, GUMMO, HOPED, MURINE, NATURE, NORTH, OASIS, PAINT, PARTY, PINNIPED, PORTRAY, PUTTY, SHIRTS, STALL, TOKEN, TRUMPS, VENOM

166. Missing word

Suggest a word to fill in the blank:

RUNWAY LINEAR

?????? WISELY

167. Enigma Variations III

Fill in the gaps such that once complete each line will in turn lead somehow to the letters E, N, I, G, M, A.

(a) The Athens band is _ _ _ .

(b) The clothes retailer is _ _ _ _ .

(c) The deductible VAT is _ _ _ _ _ .

(d) The _ _ _ _ islands are part of Nagasaki Prefecture.

(e) <U+1D110> denotes a _ _ _ _ _ .

(f) The Paul McCartney album is _ _ _ .

168. Identify the following

(a) THNH (1984)

(b) MOPI (1991)

(c) TTCSM (1974)

(d) O (1955)

(e) MB (1988)

(f) TKFM (1977)

(g) TPOP (1988)

(h) BW (2002)

169. Numerical order II

Put each of the sets of 5 words or names into numerical order:

(a) Female, Firefly, Frankenstein, Tom, Treble

(b) Beowulf, Error, Nissan, Pepsi, Pinyin

(c) Climate, Duality, Empathy, Litigate, Saturn

(d) Bit, Boston, Glaucoma, Hirudin, Wombat

170. Football logic puzzle IV

In the following table every pair of teams has played each other once. The table is ordered in the usual way (3 for a win, 1 for a draw) with goal difference, goals for and fair play record used as tiebreakers. Using arithmetic and logic, fill in all the gaps.

Team	Points	Goals For	Goals Against	Goal Difference
Rovers		1		
City				
Wanderers				
United	0			

City	v	United
Rovers	v	Wanderers
Wanderers	v	United
Rovers	v	City
Wanderers	v	City
United	v	Rovers

171. Missing

What is missing from:

BA, DWN, PLAF, AMET, NDR, ESTIN, N

172. Divide into pairs VII

Pair the following:

COBRA, EMU, GEM, LUV, MICRO, MILD, MIME, UNI, UNION, UVULAR

173. Odd one out V

Which is the odd word out?

LOGFIRE, OUTHOUSE, PENNILESS, SERPENTINE, TIMBER, TREATMENT, TUMBLEWEED

174. Main franchise

Paper towels from Stephenie Meyer

An instrument panel from Ritchie Blackmore's band

A car tool from Tim Cook

A pastry from a Graham Greene character

Which two are missing?

175. Clock II

We have previously mentioned the clock in the puzzle archive and how someone has been repeatedly sticking things on it that we have had to remove. Our first clock was damaged beyond repair and regrettably our second clock (see Tiebreaker: question 11) was too. So we bought a third clock.

This clock is again a normal clock, but instead of the numerals 1 to 12 around the clockface, it has the Roman numerals I to XII. However it would seem that this was seen as a challenge by our vandal, as they have yet again started sticking things over the numerals to give us setters puzzles to solve. As before, the 'words' or numbers form a sequence going around the clockface which indicate a time, and the hands of the clock show this time. In each case below what time showed on the clock?

(a) AS,KC,FO,OR,EM,YB,HT,VE,SI,GI,TO,SH

(b) U, IF, UJN, FS, T, BJ, XUI, WFFG, JP, D, IG, YVS

(c) K, NN, VVV, CP, Z, IV, XKK, ANNN, VG, P, VK, APP

(d) SNEEZE, LOMBOK, TURKEY, OTTERS, IVANKA, INMATE,
     MITTEN, NAGOYA, WAXING, TSETSE, VOWELS, NUZZLE

(e) VII, XIV, XVIII V'IVVII, IVXI V'IVVII IVII,
     IVXI, IVX, IIXII V'IVVII, IVIVX V'IVVII IVII,
     IVVV VIVII III VVI IVIIXI, XIVXII IVIIIVI, VVX,
     IVII?

Puzzle Hunt

These next pages contain something
a little different. A Puzzle Hunt is a
collection of puzzles, typically without
instructions. Figuring out how the puzzle
works is in itself part of the puzzle!

Each year GCHQ staff gather together in
teams of up to four, to try and solve all
the puzzles within two and a half hours.

The answers to the first 10 puzzles are all
words or short phrases.

The final puzzle is called a Meta Puzzle.
Meta Puzzles typically require you to
have solved all or most of the previous
questions before they can be tackled.

Good luck!

A. Clara Who

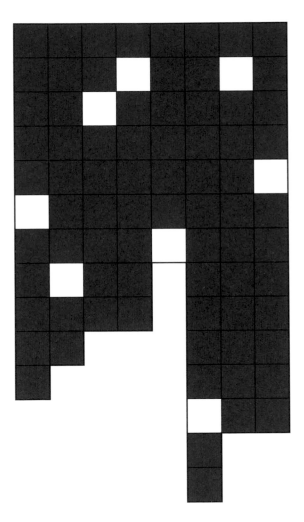

_____ _____
Travelled with her husband
But it was never a bore
With their son-in-law

_____ _____
Impossible girled
Her timeline she fractures
To help all of the Doctors

_____ _____
Her wedding didn't go well
She joined the Doctor's side
As the runaway bride

_____ _____
Escaped death's darkness
An immortal adventurer he
Was from the 51st century

_____ _____
Was Mels all along
As a way to greet she
Would say "Hello Sweetie"

_____ _____
Series reboot definer
Until she fell
Into a universe parallel

_____ _____
Had many adventures with
The robot K-9
As they travelled through time

_____ _____
The first companion
They really oughta
Explain his granddaughter

B. Spot the difference

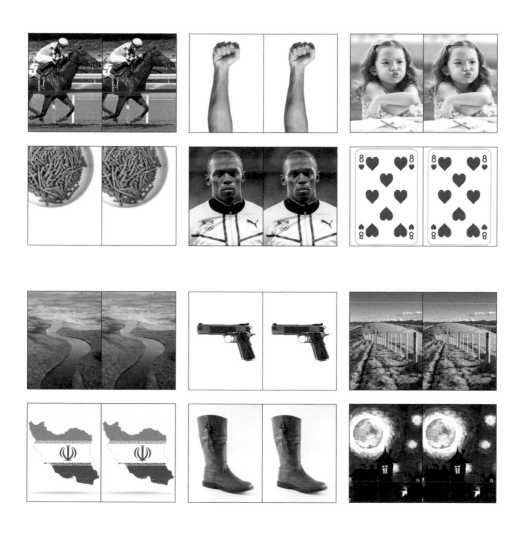

C. Chocozuma's Revenge

It's well known the Aztecs invented chocolate, but less known are the chocolate breakfast dishes they made

Turns milk
ambassadorial!

Turns milk
really quiet!

Makes milk
party!

Makes milk
merry!

Turns milk
bad!

Turns milk nut
crackingly sweet!

Turns milk
fruity!

Stops milk from
spilling!

Makes milk
chuckle!

D. Alchemy

My experiments fusing fractions of different elements together has led to some interesting new discoveries. On balance though, maybe these new creations have been found previously...

OXYPHORUS = ◯ ◯ _ _ _ _ _ _ _ _

MANGARYLLIUM = ◯ ◯ _ _ _ _ _ _ _

GOMONY = _ ◯ _ _ _ _ ◯

POLONMARIUM = _ ◯ _ ◯ _ _ _ _ _ _ _

RUTHERFROMIUM = ◯ _ _ _ ◯ _ _ _

GALLIRYPTON = _ _ ◯ _ _ ◯

HEEL = ◯ _ ◯ _

CARORON = _ _ ◯ _ _ _ ◯ _

MAGBALT = _ ◯ ◯ _ _ _ _ _

CALCUTH = ◯ _ _ ◯ _ _

E. Resurrection

New directions in genetics research has led to the feasibility of cloning almost any genome. Recreate the sequence of 80 base pairs from which these strands of nucleotides have each degraded. You may wish to carefully examine the twelve strands in their present orientations before recreating the sequence (bearing in mind common conventions such as A=1 etc.)

Reveal the next creature to benefit from the advances of tomorrow, today!

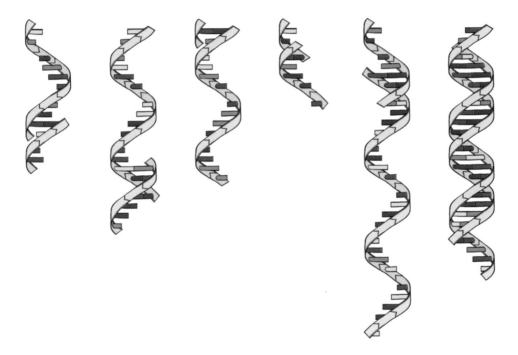

F. Kiwiana

A word that has the meaning "ocean"

Maori dance to daunt and inspire

Opalescent shell from rounded mollusc

Under the figures of NZ's coat of arms (granted by King George in August 1911)

Nephrite jade, held in high regard

Tasty dessert Australians also claim

Apparently imported from Japan, substantially

It originated within China, not down under

Not really known at all underwater, much more a confectionary aisle thing

CHOCOLATE FISH

GREENSTONE

HAKA

JANDALS

KIWIFRUIT

MOANA

PAUA

PAVLOVA

SILVER FERN

DETECTIVES

Do NOT leave password hints stuck to your computers! It will leak the master password for the entire network. I've collected the following notes on the monitors in the past week alone. Here is a list of the main offenders:

John R	5
Veronica M	6
Endeavour M	4
Hercule P	5
Sherlock H	4
Jessica F	7

Please do not let me find you doing this again!

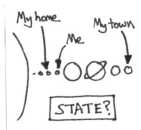

He STaREd morosely At THe BLoTTiNg paper. "it's JUST nOt my SoRToF case, lewis."

...

H. Boats and Bridges

Connect all the islands (represented by circles) into a single interconnected group. The number in the circle represents the number of bridges that connect that island to other islands. Bridges can be created horizontally and vertically, with no more than two bridges between any pair of islands. Bridges cannot cross the path of any other bridges.

Ten ships are hidden in the grid consisting of one battleship (four grid cells in length), two cruisers (three cells each), three destroyers (two cells each) and four submarines (one cell each). The ships may be placed horizontally or vertically, and no two ships can occupy adjacent grid cells, not even diagonally. The ships cannot go underneath the bridges. The numbers on the grid perimeter indicate the number of occupied cells in the corresponding rows and columns.

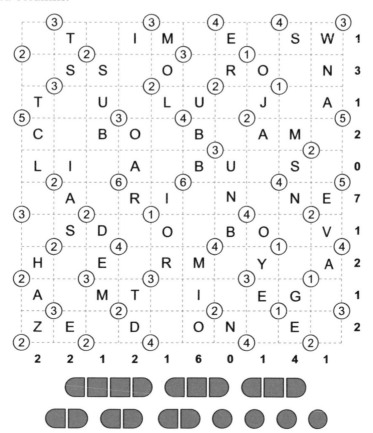

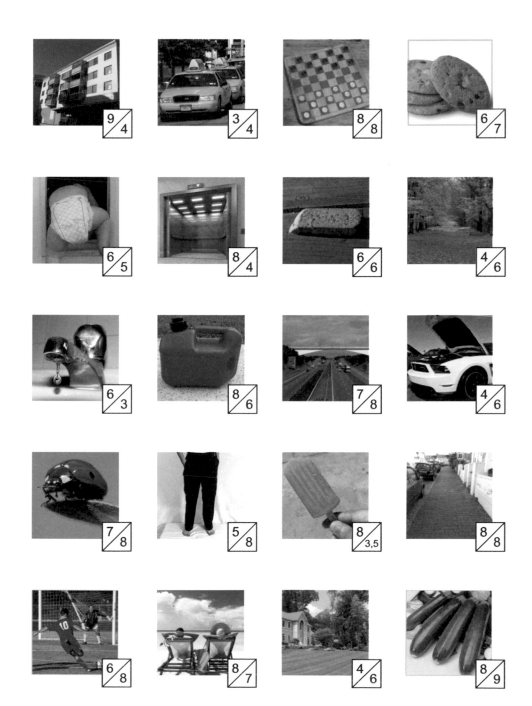

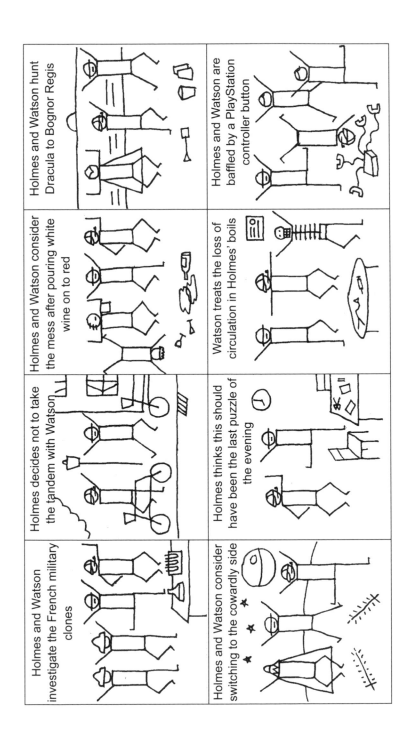

Holmes and Watson hunt Dracula to Bognor Regis

Holmes and Watson are baffled by a PlayStation controller button

Holmes and Watson consider the mess after pouring white wine on to red

Watson treats the loss of circulation in Holmes' boils

Holmes decides not to take the tandem with Watson

Holmes thinks this should have been the last puzzle of the evening

Holmes and Watson investigate the French military clones

Holmes and Watson consider switching to the cowardly side

Meta Puzzle

Use your previous answers to fill in the grid and reveal a message

					3				D			
	6						3					
							J					2
	6											
	G				3			7	H			3
		2							A			
			I		2							
						13				1		
C												
									2			
	18						13					
B				10								
18	3		10				E			2		
						F						

21 green islands can be found in a grey sea.
Islands are formed of a contiguous block of connected cells.
Some island cells contain numbers revealing the size of the island.
Islands are surrounded by sea on all borders, but may touch diagonally.
The sea is all connected.
There are no 3x3 squares containing only sea.

If you place the answers to the preceding puzzles into the grid then:
The vowels are all in sea.
The consonants are all on islands.

Einstein's Aphorism

Einstein's Aphorism (6, 5, 8)

This was part of a set of puzzles which explains the slightly odd answer!

amass+priors=anoint

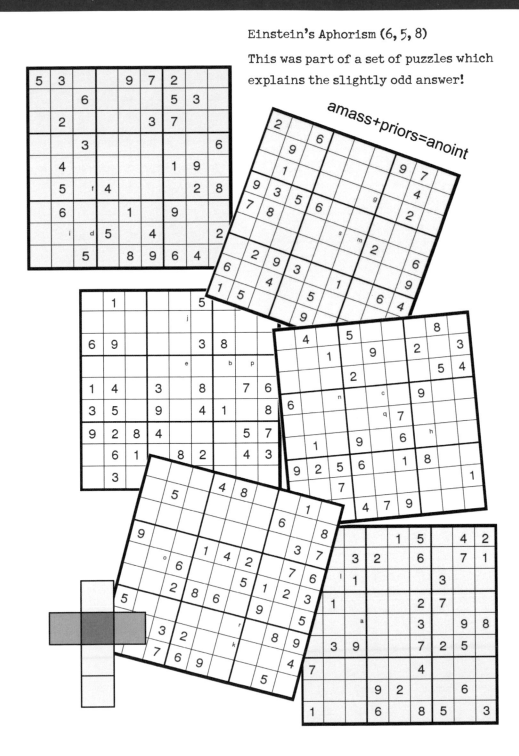

Dingbatagrams

(a)

(b)

(c) 0x47

(d)

(e)

Sayings

In what sort of state might I end up?

(a)

(b)

(c)

(d)

(e)

(f)

(g)

(h)

(i)

(j)

(k)

(l)

(m)

(n)

(o)

(p)

(q)

(r)

(s)

(t)

(u)

(v)

176. Motorway darts

Motorway darts is a popular game in GCHQ. You throw darts at a map of the UK and score points corresponding to the name of any motorway you hit. The aim is to score 501 points. For example, you could throw 501 darts at the M1, or 20 darts at the M25 and one at the M1.

Can it be done in 3 darts, and if so, how?

177. Also

Identify:

Largest marsupial
Wash the scalp
Military display
Snow home
Indian potato

What are we celebrating that looks like it belongs with the set above but doesn't?

178. Identify the following

(a) Heiress who built a rambling mansion in San Jose following the advice of a medium

(b) Toronto subway station

(c) TV historian, born in Manchester in 1948

(d) Metonymic road in London

(e) Tolkien character – son of Malach, father of Hathol

(f) Actress, born in 1977, daughter of a Dame

(g) Thin-leafed deciduous hardwood tree

(h) British celebrity chef, born in 1966, whose restaurants have been awarded numerous Michelin stars

179. Alphabetical order

Put these into alphabetical order:

Wells
Washington
Tellytubby
Sky Airlines
Qualification
Original Starfighter
Nevermind
Lunar Lander
-k
Iron
Eliot
Déjà
CO_2/O_2

180. 12 things

What completes:

ar, ar, m, ne, l, g, br, coro, r, cr, ?, ?

181. Interesting numbers

(a) What is the fourth and final member of this set: 19, 28, 81, ?

(b) Similarly, and ignoring 1, 2 and 6 as trivial, what is the sixth and final member of this set: 7, 22, 38, 56, 101, ?

(c) 1098003 is the lowest number with what property?

182. Wicked

Which is the odd one out?

ALTERNATIVE, BENIN, PANNIER, PARIAH, PARTNER, STACKABLE

183. Perhaps there is a key

Where?

(MEN, (MAY, (LED, RAID), TYRES), AGAIN)

184. Word divisions

What comes next?

0, 25, 96, 23, 176, 84, 20, 57, ?

THE GEOGRAPHY OF BRITISH SIGNALS INTELLIGENCE

GCHQ's primary focus has been on foreign communications to identify threats to the UK. For a century, British Signals Intelligence has had to adapt its approach to the ever-developing technologies of communications and this has included using a variety of locations.

Signals Intelligence doesn't begin with an individual transmission being intercepted: it begins with a broad understanding of how communications systems work. An understanding of the communications technologies available to the individual of interest is then overlaid. The details of the specific system the individual of interest is using is then applied. But even that knowledge isn't enough: unless a method of collecting that system is in the right place, the target's communication cannot be intercepted.

Sometimes the right place is easy to identify: when GCHQ realized in the 1960s that it would have to start intercepting communications being relayed via satellites, it was clear that the only part of the UK from which both Atlantic Ocean and Indian Ocean satellite transmissions

could be seen was Cornwall. The decision to open a new
station outside Bude was easy. Collection of high-frequency
shortwave signals was constrained in a different way. What
was needed was enough land to construct both a station
and the appropriate antennae, and in an area with a low
'noise floor': usually a semi-rural area away from local
transmissions that might swamp those of the targets.
Direction-finding (DF) stations need to be built along
a baseline. In this way, it is possible to calculate the
direction from which intercepted signals are coming, and
from as varied a set of locations as possible.

Non-technical factors could also play a part: from the
inter-war period to the Cold War, when the three service
ministries (the Admiralty, the War Office and the Air
Ministry) were responsible for their own interception
sites, there was no expectation that they should share
capabilities. As a result they each built their own
collection and DF stations where it suited them (and the
Americans, Canadians, Free French and Poles all built
their own facilities in the UK as well during the Second
World War). All these sites needed to be connected to each
other, and to other facilities that needed to be closer to
the recipients of the intelligence product, all of which
required an extensive telecommunications infrastructure.

As a result, around 200 separate sites have been used in
some way to enable intelligence collection or production
since August 1914, with most of them supporting the
military during the Second World War and the Cold War.
Bower, Brora, Bude, Cheadle, Culmhead, Digby, Flowerdown,
Hawklaw, Newton and Tean have all played a part. We know
lots about some of them, and very little about others. Some

have disappeared entirely, leaving no trace at all; while others, long since abandoned, have left very visible traces on the landscape. Only one, GCHQ's station in Scarborough, has been open throughout our history (in fact, it predates the formation of GCHQ under its original name of the Government Code and Cypher School in 1919).

Changes in communications technology have removed geography as the decisive factor in the interception of communications, and the same technology has facilitated people working together even though they aren't in the same place. The number of sites today is much smaller as a result.

185. Glass hills

Looking from the direction indicated, spell out these historic sites and see what you discover:

BUDE	NW
TEAN	W
DIGBY	SE
NEWTON	SE
NEWTON	SW
BUDE	NE
CULMHEAD	E
BOWER	SE

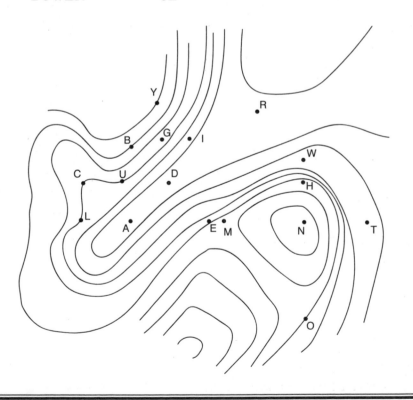

186.   Four words

I have written a word. I cross out the first letter, and add a letter to the end to make a new word. Then I do this a second time, and a third. For example, I could have written the words SHOP, HOPE, OPEN, PEND. However, with the words I've actually written, the first pair sound the same, as do the last pair. What are the four words?

187.   **Which?**

The following list of 55 numbers can be divided into 10 sets, all of different sizes. Put another way, there is one set of 10, one set of 9, etc. Which number is in the set of one?

6	8	13	14	15
17	18	19	21	24
27	37	40	49	56
120	144	160	200	273
372	433	519	721	732
925	946	1134	6121	8011
9627	9729	11235	12481	24681
38809	51197	123211	205030	823543
948437	1000006	1330210	1594323	1652158
4002103	9447539	94478533	96335533	743665127
995674663	34159739501	37907741687	8000000000085	5559060566555523

188.   **Bond, James Bond**

What completes?

GoldenEye

Casino Royale

The Living Daylights

You Only Live Twice

On Her Majesty's Secret Service

?

189. Sequence III

In (c), the sequence is completed by a biblical location.

(a) Cimarron, –, Hamlet, All About Eve, The Godfather, The Deer Hunter, Platoon, Unforgiven, The English Patient, ?, ?, ?, ?

(b) 206, 230, 250, 260, 602, 640, 5000, 8000, 9000, ?, ?, ?

(c) Sea, Hay, Char, Rye, Yes, Tea, ?

190. Second in line

Who is currently last in the sequence?

Anne, Elizabeth, Alexander, Grace, Elizabeth, Arthur, ?

191. Fun with flags

Which UN country has a flag with no colours in common with the Union Jack (i.e. it has no shade of red, white or blue)?

192. Relationships

(a) WRY is to MAURIE

as PEEVES is to which national animal?

(b) MCCARTNEY is to MANILOW

as TOME is to ?

(c) SANTA is to DECEMBER

as FLAPJACK is to what variety of mango?

193. Reorder I

Find a connection between the following items and reorder them:

AUDIOTAPE, CAPSIZE, HOME GAME, MOTHERHOOD

194. Century mason

What links the answers to these questions?

(a) The emblems, symbols, or paraphernalia indicative of royalty

(b) An imperial unit of length

(c) A genus of snake named after Australian politician William Denison

(d) A weather phenomenon

(e) A type of port

(f) One of the General Synod of the Church of England's houses

(g) One of the Gallagher brothers

(h) This word from the first verse of the National Anthem is a homophone of the answer to question (d)

(i) An album by the Lightning Seeds

(j) The Spanish word for 'angels'

(k) A viral disease carried by bats and other mammals

(l) A model of Toyota car

(m) A dance that is also a word in the NATO phonetic alphabet

195. Hidden mathematicians

Which mathematicians are hiding in the following sequences?

(a) 8, 34, 1, 610, 377, 1, 2, 2, 34

(b) 136, 1, 190, 6, 1, 78

And, finally, whose elements are these?

(c) Europium, Chlorine, Iodine, Darmstadtium

196. Lottery

What is Charmaine's lottery entry? (A lottery entry is 6 different numbers in the range 1–49; they are customarily displayed in ascending order.)

```
Anthony        8 11 14 15 25 42

Charmaine      ?  ?  ?  ?  ?  ?

Dorothy        2  3  8 25 32 41

Geraldine     11 18 24 26 42 45

Justine       10 19 20 21 42 45

Orville       15 18 21 22 25 42
```

197. Not a mistake

Whats the connection between:

Halloween, Hawaii, Oumuamua, Quran, Tiananmen

198. Properties II

Each of the words to the left of the colon has a property that the word to the right of the colon doesn't have. What is the property?

(a) AVOCADO, BALUSTRADE, CLOTHO, DATABASE, ENTENTE, FABULOUS, GEOGRAPHIC, HOMESICK, INQUISITOR, JACARANDA : KALAHARI

(b) AVOCADO, BALUSTRADE, CLOTHO, DATABASE, ENTENTE, FABULOUS, GEOGRAPHIC, HOMESICK, INQUISITOR, JACARANDA : KLONDIKE

(c) AVOCADO, BALUSTRADE, CLOTHO, DATABASE, ENTENTE, FABULOUS, GEOGRAPHIC, HOMESICK, INQUISITOR, JACARANDA : KOSOVO

199. Question 199

Complete the table:

Bohemian Rhapsody	Love & Pride
Say I'm Your Number One	?

200. Enigma Variations IV

Fill in the gaps such that once complete each line will in turn lead somehow to the letters E, N, I, G, M, A.

(a) The composer wrote the _ _ _ _ _ _ _ _ _ _ _ _ _ _ _ _ _ .

(b) The comms officer was Lt. _ _ _ _ _ _ .

(c) The Tsar was known as the _ _ _ _ _ _ _ _ .

(d) The actress wanted to be _ _ _ _ _ .

(e) The punk innovator managed the _ _ _ _ _ _ _ _ _ _ .

(f) The stadium hosts the _ _ _ _ _ _ final.

201. Identify and divide into pairs

(a) Album by Mark Lanegan

(b) Derbyshire village

(c) Oxfordshire village

(d) Naval version of 'wilco'

(e) Mary, Queen of Scots was held here for six months

(f) Won 25 caps for the Wales national rugby union team

(g) Village near (e)

(h) W. H. Auden poem

(i) Lemur

(j) Bond composer

202. No words

(a) 83192 71236 77119 11675 97326 11161 47477 14713 71191
11143 54772 34317 47137 11961 11115 34723 43714 74311
13477 36147 43111 32379 11432 34311 11715?

(b) 13773 41813 42333 41441 25841 44750 25286 57211 67653
44181 67652 15537 72610 33437 71361 08610 37759 87610
34377 67654 18134 46368 61037 75534 13216 76512 13935
25846 10676 52125 84555 67652?

203. Reorder II

Reorder: ARCH, AY, EMBER, GUST

204. What?

What are?

BZL	CTJEZBSZ	DZRBTDFBS	DZRBTDFBSA	DZSDAB	DZSDAB
EBAACK	EBZCKZS	FVAS	GAAMQ	GAF	GAHFSAK
GTFS	GTOBZ	HMSNCBMS	JDFBRTF	JDFBRTFS	JDUFBRTFAS
JEFBRTFSAS	JEXKKAS	JKAASOFE	JKASOFDEAS	JKTAB	JZNTKKZBTMJ
JZNTKKZTBA	KPTGGTSNZBSZ	KQAAGTSNAS	KZMBAZM	KZMBMJ	LZZNI
NALAZMV	NALTST	OAGTAB	OFFNJDUMKKAB	OZGZSDA	PABJAZM
PTABNA	PTBNF	PTJJAS	PZIMBAS	PZKKMLZSSAS	PZNAS
QAANJDUZZG	QZKABLZS	RFTJJFSJ	RTJDAJ	ZBTAJ	ZWMZBTMJ

205. Neighbours with nothing in common

Name two neighbouring European countries (i.e. sharing a land border) whose English names have no letters in common.

206. A particular question

Group the following words into sets of 3:

AGE, ARM, BASE, BREAK, CHAR, CUPOLA, FATHOM, GEAR, GUN, HARM, LUPINE, LOWEST, MARKS, RAG, SAGE, SHOW, SUPPORT, TOUCH

207. (2,3,101,2)

(2,3,101,2) (1,4,1000000,1,3,101,11,6) (6,3,1,4,11,100) (6,2,101,1,100) (4,1,3) (2,3,1) (11,1,2,2,1,3,6) (5,1) (2,3,1) (101,1,6,2,1,3) (2,1) (4,1,4,3) (2,5,1000000,1,6) (4,5,5,1)?

208. Divide into pairs VIII

Wear? Pear...

AL, ANNA, BAN, BAY, BULL, CAR, CAR, HAVE, JEERS, JOULE, KEY, LEA, LOW, MAR, MAY, ROUTE, SORE, TOE, TOMB, WORE

209. Connexions

0 π^2
1 $\sqrt{2}$
2 $y\sqrt{y}$
3 2π
4 $\pi + e$
5 π
6 $\sqrt{5}$
7 $n\sqrt{3}$
8 e
9 $\sqrt{5} + 2\sqrt{3}$
10 $\sqrt{2} + 3e + \pi$
11 $8\sqrt{l}$
12 $\pi + 5\sqrt{2}$
13 $3\pi + 6\sqrt{2}$
14 $6\sqrt{2} + 4\sqrt{3}$
15 $2\pi + z\sqrt{2}$
16 $\sqrt{5} + e + 5\sqrt{3}$
17 $2\pi + 9\sqrt{3}$
18 $2\pi + 7e$
19 $x\pi$
20 $\sqrt{2} + e + m\sqrt{3}$

$x\sqrt{(lmyz)} + n\,(m + l) + l^2\,(l + z)$

NCSC – SECURITY IN THE TWENTY-FIRST CENTURY

GCHQ's intelligence mission is only one side of the story. The complementary mission of the defence of the UK's secrets is just as vital and today is carried out by the part of GCHQ known as the National Cyber Security Centre (NCSC).

In the form of CESG (the Communications-Electronics Security Group), GCHQ had historically advised on the mathematical security around the most important UK government communications. As the internet age dawned, it became apparent that guidance was also needed to make sure that there were no vulnerabilities in the software that implemented the mathematics, the operating system that accommodated the software and the hardware that ran the operating system. Cryptographers now needed expertise in a wide range of disciplines. A new breed of security architect arose, able to assess and address the security of whole systems.

At the same time the technical landscape was changing, the number of users entering it was growing. Secure communications were no longer just the concern of armed forces and diplomats. Within government, more and more groups were in search of advice; GCHQ's technical strength made it the obvious port of call. The security mission at GCHQ wrestled with increasing demand. Five areas of focus were identified:

- Providing a high level of assurance to the UK's most sensitive communications.

- The government needed to run its own IT departments and networks as an exemplar of good security.

- The government as a provider of online services (e.g. car tax, self-assessment income tax) needed to make sure that those services were secure for both government and users.

- The government needed to be able to defend critical national infrastructure (power, water, transport) from online attack.

- Most ambitiously of all, GCHQ wanted to make the UK the safest place to live and do business online.

These new challenges required a different approach. GCHQ could not demand absolute adherence to approved security practices, as it did with high-threat customers, nor was the highest-grade security appropriate to all threats – or affordable for many organizations. Instead, users needed to understand that absolute security is a myth and that decisions and trade-offs must be made between usability and risk. And if they outsourced their system to another organization, they needed to be able to make informed decisions as to who they could trust with it. The security mission now needed to operate in a more open way with organizations and the public, using influence and advice rather than decree. GCHQ could not hope to raise each manager or user to its own National Technical Authority level, but it could aim to make it much easier for them to make better security decisions.

GCHQ recognized it needed help with a challenge on this scale. It had to reach out to industry and academia. To an organization that was used to living and working in secrecy, openly discussing its work was a major cultural shift. Programmes such as the Academic Centres of Excellence (working with higher-education establishments) and the Cyber Accelerator (helping to grow start-ups in cyber security) drove the new partnerships forward, but GCHQ knew this wouldn't have the enduring impact it needed. In order to make sure that the UK had the skills for the future, initiatives such as the CyberFirst Girls competition – with entries from thousands of thirteen- to fifteen-year-olds – sought to develop the skills pipeline and increase diversity in the cyber security industry.

GCHQ was starting to change the online security of the UK for the better. Simple, usable advice such as the 'Ten Steps to Cyber Security' made it easier for UK businesses to secure their networks. In October 2016 the security mission took its greatest step into the open with the creation of the public-facing NCSC, as other teams from across government joined their GCHQ colleagues to centralize the national expertise on cyber security.

The distinctive red glass Nova South building in London is now home to a world-class technical organization able to provide both radically new security thinking, such as its user-focused practical password advice, and rapid response to online threats such as the WannaCry attack.

In its first year of operation, NCSC responded to 590 significant cyber attacks which included targets such as the NHS, and the UK and Scottish parliaments. The NCSC Active Cyber Defence Programme has reduced the average lifespan of a fake phishing site from twenty-seven hours to under one hour.

210. Domino logic

We can use dominos to create logic gates; you just have to work out how the dominoes will fall.

Pushing a domino input represents a 1 on a logic gate. Leaving a domino standing represents a 0.

Can you work out what would happen when you push the dominoes and use it to complete the logic tables below?

Domino Logic 1

IN A	IN B	OUT C
0	0	
0	1	
1	0	
1	1	

Domino Logic 2

IN A	IN B	OUT C
0	0	
0	1	
1	0	
1	1	

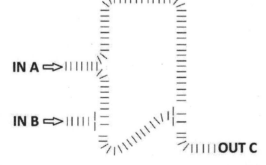

Domino Logic 3

IN A	IN B	OUT C
0	0	
0	1	
1	0	
1	1	

211. Lumber

Arrange in a different logical order:

FIX, JIVE, NIFTY, NIGHT, PEN, PINE, SHIRTY, SOUR

212. Divide into pairs IX

Divide into pairs:

(a) AVAR, BPZMM, CQRACNNW, HLJKW, JMZI, KTZW, OKPWU VYQ, RUM, UFO, YXO

(b) ASxUBE, FDHVDU, FLSKHU, IIDNPONLTOSTOR, KJDYPFKAPUAQTB, KRIKQTQRQTFD, KTCGAL, LAYF, PTxAOIxNSOx, PYRB, QROPRCRRRCJH, TSINR

213. Roger that

Fill in the gap:

Snow	White
Cool	Blue
Small	Red
Clever	Orange
Clumsy	Green
Impossible	?

214. Roman numeral sums

Some positive integers, such as five, contain Roman numerals. The sum of the roman numerals in five is I+V = 6. Eleven has a sum L+V=55, which is nice because 55 is a multiple of eleven. What other positive integers have this property? (We believe there are six others.)

215. As easy as A, B, E, ...

What comes next in each of these sequences?

(a) A B E Z H I K M N O P T Y ?

(b) A B E K M H O P C ?

216. Sequence IV

(a) 111, 221, 23311, 4143, 4551, 656, 61511, ?, ...

(b) 250, 2660, 450, ?, 200, 550, 2660, ?, 600, 50, 750, ...

(c) 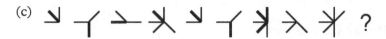 ?

217. Next in succession

What comes next in succession?

W, I, H, S, E, R, J, N, E, D, W, I, R, ?

218. Reorder III

 Find a connection between the following items and reorder them:

JACKSONVILLE, LIECHTENSTEIN, MACEDONIA,
QUEENSLAND, UNITED KINGDOM

219. Some numbers

(a) What is the next number in this sequence:

–,–,1,3,8,13,14,25,24,74,105,116,127,?

(b) Fill in the gaps:

0, 0, 0, 0, 40000000000000000000000000000, 11, 13, 72, 72, 72,
?, ?, ?, ?, ?, 115, 304, 304, 304, ...

220. Identify

(a) Identify these words:
3116, 4215, 6924, 7513, 161815, 6181513

(b) Identify these different words:
3116, 4215, 6924, 7513, 161815, 6181513

221. Enigma Variations V

Fill in the gaps such that once complete each line will in turn lead somehow to the letters E, N, I, G, M, A.

(a) The American beer is _ _ _ _ _ _ _   _ _ _ _ .

(b) The Highway ends in _ _ _ _ _   _ _ _ _ _ _ .

(c) Internet pagers use the _ _ _ _ _ _   _ _ _ _ _ _ _
_ _ _ _ _   _ _ _ _ _ _ _ _ .

(d) The album is by _ _ _   _ _ _ _ _ _ _ _ _ _ .

(e) Danger Man was followed by _ _ _   _ _ _ _ _ _ _ _ .

(f) The Bond villain was played by _ _ _ _ _ _   _ _ _ _ .

222. A musical interlude

Which song or piece of music?

(a) 711334814/17491817

(b) OVFZ ADQZ

(c) D'ARBANVILLE, ELEANOR, MADONNA

(d) A NMBER ONE HIT SINGLE

(e) THREE MAY HESITATE

(f) A or C or E or I or M or R

(g) MMEETHUSSSELAAGEH

223. _ _ ' s   _ _ l _ _ _ _ k _ o _ e

What comes next:

A, A, A, A, N, A, A, A, A, A, A, ?, ...

224. Leopards in the barrows

(a) Oral inspection	(g) Scabrous heap
(b) Midlife sprees	(h) Pirate cap
(c) Epic dream	(i) Unique snort
(d) Irradiating columnists	(j) Anagram act
(e) Dinosaur moped	(k) Sofa topic
(f) Resonant paragon	(l) Venison trivia

225. Jigsaw

Ordinals:	1, 3, 4, 6, 7, 10
Rainbow:	1, 2, 3, 6
Cardinals:	1, 4, 5, 6, 7
Greek:	2, 4, 5, 9, 13, 14, 16, 23
Planets:	1, 2, 3, 4, 5
Popes:	12, 45, 53, 121

226. Reorder IV

Find a connection between the following items and reorder them:

LIBRA, NEED, ROMANCE, YAM

227. Three shelves

Our chemistry laboratory has a selective display of elements from the periodic table. It's arranged on 3 shelves and the top shelf contains 11 elements ranging alphabetically from ERBIUM to TITANIUM. The middle shelf has 12 elements ranging from ALUMINIUM to SILVER. What are the elements on the lower shelf?

228. Something inobvious

Which is the odd word out:

BASALTIC, COCHLEAR, ECSTASY, HAZARD, PLANTAIN

229. Identify the following

(a) Live Blondie album	$a = i + (f - j)$
(b) Croatian winner of US Open	$b = c - ((i - j) - (d - a))$
(c) South-east Asian capital	$c = i - ((f - j) / (a - i))$
(d) Former Manchester United defender	$d = a + ((i - j) - (c - b))$
(e) Italian singer's single	$e = i + ((a + j) - (c - b))$
(f) Mayor's car?	$f = j + (a - i)$
(g) Tribe	$g = e + j - (b / (a - i))$
(h) Indian village	$h = g + (i - j) + (d - a)$
(i) The Apprentice by a Scandinavian name	$i = j + (c - b) + (d - a)$
(j) Roman name for Valencian town	$j = f - (a - i)$

230. What is the odd word out?

 CHAT AS SHE OLD LORD NUT

231. Reorder V

Arrange the following words into a different logical order:

CENT, DOZE, DUE, ELF, HAT, PUMP

Which word is ONCE in the list?

CODING TODAY

The basis of Signals Intelligence has been breaking the
code protecting the secret information you want to get to.
That's still true today but, as technology has advanced, a
different type of code has also come to be of importance
to GCHQ.

Throughout its hundred-year history, GCHQ has been
synonymous with codes. It all started with code- and cipher-
breaking throughout the First and Second World Wars to
gather intelligence on the threats of the day. At the same
time, codes and ciphers were being designed to help keep
Britain's secrets, well — secret! This mission continues
to this day through the work of the National Cyber
Security Centre (NCSC), a part of GCHQ, in securing the UK's
communications and systems.

These days a different type of code lies at the heart of the
business. With the advent of computers, understanding
them and programming them has been essential to enable
GCHQ's work. The world now runs on data which underpins
decision-making by systems with artificial intelligence
that communicate only with other systems. So GCHQ needs
coding in order to build systems to make sense of that

complex data and find the valuable intelligence it needs in an increasingly noisy world of communications.

Different problems often need different approaches to solve them, and that is why there are so many coding languages out there, each suited to solving different kinds of problems or building different kinds of systems — more are emerging every year! Looking at similar systems or spending a short time researching your problem usually tells you quite quickly which coding language you should use. The approach you want to take, if you need to integrate with an existing system, and what platforms it will be running on, are usually factors too.

Beyond those who can code, GCHQ also needs mathematicians, cryptographers, linguists, analysts, security experts, systems engineers and people skilled in many other areas. They all need to work together on problems, each bringing a different way of thinking to find the best solution. Computers can only take you so far: humans and their neuro-diversity lead to ingenious thinking and this has been of constant importance throughout GCHQ's history and will undoubtedly continue to be so in the future. Great code allows GCHQ to do its job keeping the country safe from strategic national security threats, terrorists, drug traffickers, cyber-criminals and more.

232. Holiday

I've started making my holiday plans in the form of this Python function. Where am I going? Where am I flying from?

```
def fun(n):
    a,b,c,d,e = [int(x) for x in '{:05b}'.format(n)]
    p = (1^c) &(a | b) ^ (1^d)
    q = a &b &(1^c) & (1^d) ^ d & e & (a | (1^c))
    r = 1 ^ e &(b | c | d)
    s = (a ^ e) &(b ^ c ^ d & e)
    t = (1^b) ^ e ^ (d & (c | e))
    return int(''.join('01'[i] for i in [p,q,r,s,t]),2)
```

233. Gone to pot

Which of EAT, ALL or LOW could finish:

TON, HAM, BAN, LET, HOE, LOO, ?

234. AA to ZZ

AA = BOS, CAM, STNY	NN = TRM, TSW, TW
BB = DG, KB, SS	OO = APIG, G, LN
CC = C, IHON, IOL	PP = BISS, CU, L
DD = CJ, PT, TTOM	QQ = TB
EE = TBC, TMD, YG	RR = AWITW, IP, OOA
FF = HOSAF, T, U	SS = O, R, R
GG = AYDM, N, WO	TT = HF, MB, YMMLY
HH = BN, RA, TP	UU = HIAP, ML
II = ABD, SIM	VV = CR, TI, WC

JJ = CB, LIAMST, LL   WW = PITS, SBM, TLS

KK = BILB, PAP, TIG   XX = RCDN

LL = FF, HFL, MG   YY = LDT, MH, VR

MM = DBC, FSOJ, I   ZZ = CTHD, SR, ZW

235. Increasing scores

The abbreviation GCHQ has the curious property that each of its letters has a higher value in Scrabble than the previous one. Can you name an appropriate three-letter word with this property?

236. The clue is in the question

How do you find the hidden messages?

(a) DR. AWK CAB KOOL

(b) IMAGINE ORES PURIM BELLE TOTERS

(c) CARLESS CHOKE CHOKE DIABLO EMBERS EOM FACADE FROM GOD HOTEL MEOW MODEL OTTO PALINDROME PAMPA QUEBEC SIN SPA SUPERHERO TANGO TELEPORTER TITAN TYPOS UNREAL USTINOV WHO

237. Non-trivial numerical anagrams

Some positive integers, when written out in words, are anagrams of others (e.g. 'two thousand and sixteen' and 'sixteen thousand and two'). In most cases, these numbers written out in digits are also anagrams if you ignore the 0s.

Find two positive integers whose English names are anagrams but whose digits are not, even ignoring the 0s.

238. Reorder VI

Find a connection between the following items and reorder them:

DULL SENSE, NEWER SWELL, NEWS LURED, SUDDENNESS

People I

Identify the people and divide into pairs.

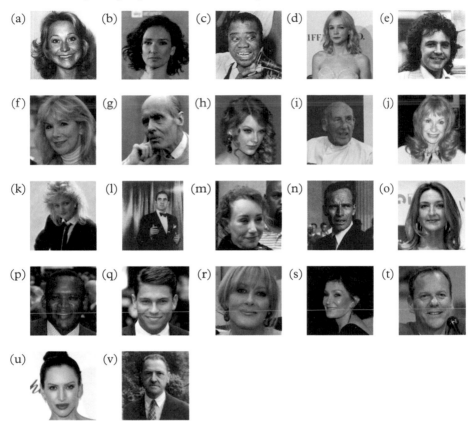

(a) (b) (c) (d) (e)
(f) (g) (h) (i) (j)
(k) (l) (m) (n) (o)
(p) (q) (r) (s) (t)
(u) (v)

Picture sums I

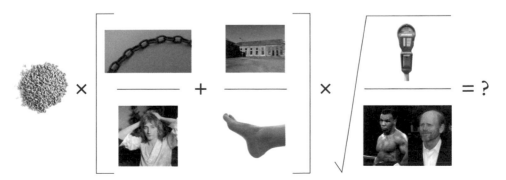

Picture sums II

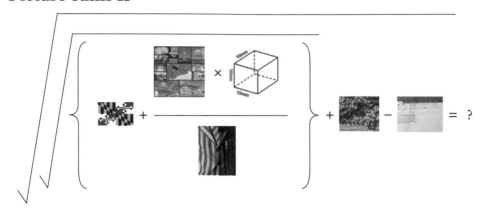

Picture sequences

(a) Identify the following people - they're in order, which should help in working out who might be first and last:

(b) Complete the sequence:

(c) Identify the following and suggest an image for the final picture:

Picture pairs I

Divide into pairs:

(a)

(b)

(c)

(d)

(e)

(f)

(g)

(h)

(i)

(j)

(k)

(l)

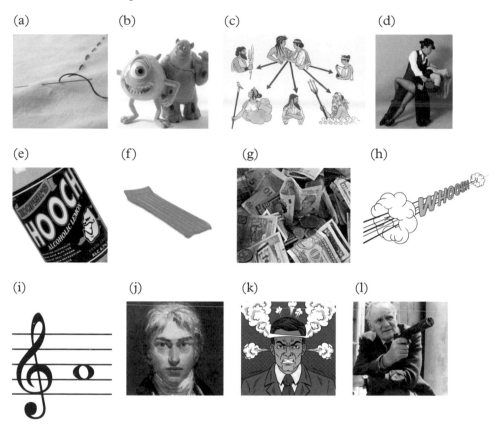

Picture connections I

Identify these people and explain the connection:

People II

Indentify these people and explain the sequence:

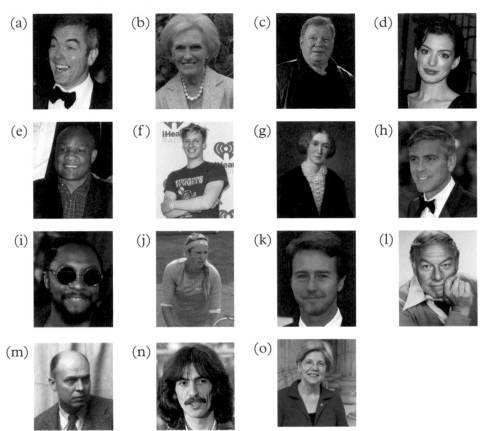

(a)   (b)   (c)   (d)

(e)   (f)   (g)   (h)

(i)   (j)   (k)   (l)

(m)   (n)   (o)

Picture sums III

Please answer with a picture of one person:

+
−

= ?

Picture sums IV

People III

Identify these people and explain the connection:

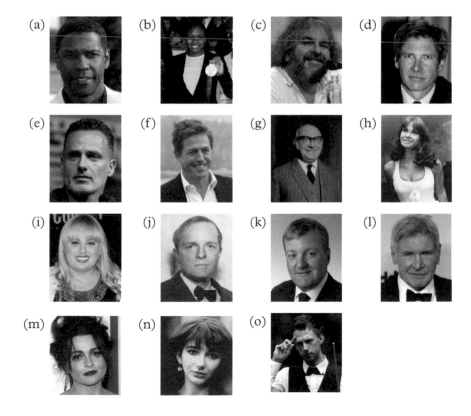

(a) (b) (c) (d)

(e) (f) (g) (h)

(i) (j) (k) (l)

(m) (n) (o)

People IV

(a) Identify these people and divide into two opposing teams:

(i)   (ii)   (iii)   (iv)   (v)   (vi)

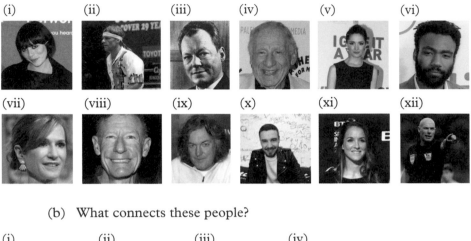

(vii)   (viii)   (ix)   (x)   (xi)   (xii)

(b) What connects these people?

(i)   (ii)   (iii)   (iv)

Picture pairs II

Divide into pairs. Which is the odd pair out?

(a)   (b)   (c)   (d)   (e)

(f)   (g)   (h)   (i)   (j)   (k)

(l)   (m)   (n)   (o)   (p)

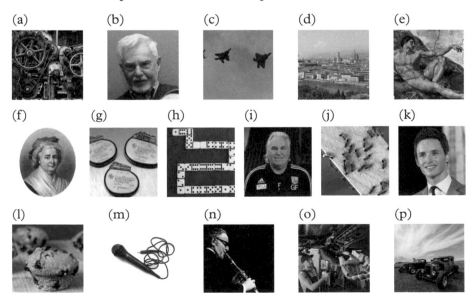

Wordsearch

The grid contains twenty strongly connected answers. For each answer you give, please also give the coordinates of its start and end cells. We used Wiktionary and Wikipedia as references.

People V

Identify these people and explain the sequence:

(a)  (b) (c) (d)

(e) (f) (g) (h)

(i) (j) (k) (l)

(m)

239. **Identify the following**

(a) Indie band from NY state

(b) US (or maybe NZ) tennis player

(c) 1995 Michael Jackson single

(d) TV series starring Kristen Bell

(e) Character in Watchmen

(f) Videogame console

(g) District of Bucharest

(h) WWII fundraising pig

240. **Enigma Variations VI**

Fill in the gaps such that once complete each line will in turn lead somehow to the letters E, N, I, G, M, A.

(a) Novocaine is a _ _ _ _ _ _.

(b) The abbreviation for synchronisation is _ _ _ _.

(c) AIBO is a _ _ _ _ _.

(d) The rate of change of momentum is proportional to _ _ _ _ _.

(e) The square is _ _ _ _ _ _ _ _ _ _.

(f) The tributary of the Tyne is the _ _ _ _.

241. **National numbers**

Canada = 1.7

USA = 4.7

Australia = 26.1

France = ?

242. Tournament

Which wins?

SWING *v* ESSAY   _____

CUTS *v* MAIL   _____   _____

                                     _____

DEED *v* HOLD   _____   _____

MUSIC *v* SNIPS   _____

                                           _____

RAG *v* TOR   _____

SHEARS *v* THESIS   _____   _____

                                     _____

MIRROR *v* COBBLE   _____   _____

ZIGZAGS *v* DISTURB   _____

243. The table

Each item in the following table has been shifted by its 'value'. Good luck identifying the missing entry.

??	DG	IJ	KK	KY	U
DG	LJ	KK	KY	U	MM
IJ	KK	NY	U	MM	W
KK	KY	U	PM	W	OD
KY	U	MM	W	RD	J
U	MM	W	OD	J	N

244. Sequence V

(a) 9780319243664, 9780319243671, 9780319243688, 9780319243695, 9780319243701, 9780319243718, ?

(b) BAR (13), LOG (8), PANDA (4), DRAG (14), BEAR (4), TOE (4), FRIAR (13), ?

(c) How does this number continue? Like its more famous cousin, it continues forever, but you need only provide the next 5 digits:

5.343443345454545?????...

245. Some odd characters

Identify the odd one out:

BOB ROSS, KATY HILL, KENNETH TYNAN, TAMMY WYNETTE, ZAYN MALIK

246. A balanced clock

Rearrange the numbers I to XII around a clock face such that the centre of mass of the clockface is in the centre. You may assume that the centre of mass of each number is in the centre of that number, and that the mass of X = the mass of V = twice the mass of I.

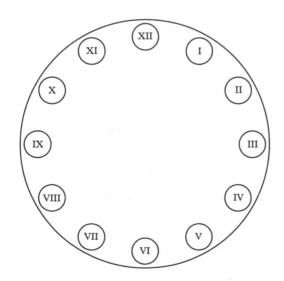

247. Seasonal greeting

A key stream consisting of a repeating 3-letter word has been added (A=1, B=2, etc) to a list of associated words. Once deciphered, extract the seasonal greeting from the setters.

bdeijds hmgaal aahn ddbqwrb wnzcj kpaddf

odlacdf wnxxltghejv ddbqbjdn rbtlmr bbgdot cmgamvxz
bdslxdi jzxbq hjehjrgje

248. Speak for yourself

Who are: DL&P, KH&O, RA&LC, RDC&NB, SL&LC

249. What connects?

What connects Elkins, Fairchild, Heller, Kovic, MacGuff, Poulain, Wormwood?

250. Where on earth?

66686668668666866666262268622662262666666686688688686
88868888484884848844888488488488422224224484484886684
44484244224248442424242422424424444224222622262262222

251. Where in the world?

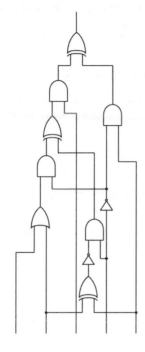

252. Mr mssng vwls

Wht's nxt n th sqnc?

(a) M, N, X, ?

(b) CHRC, SRKZY, HLLND, ?

(c) NGSTRM, NNMTR, TN NNMTRS, N HNDRD NNMTRS, ?

Whch sng?

Whch pngn?

253. And finally …

Which of the following words – AT, HERO, RAVE, SAKE – could complete this sequence?

WAS, BAR, SAIL, TIER, LOSE, PLACE, ?

Entrance Tests

254. 1968 Entrance Test

1.   Background

The hall in which the local puzzle club meets contains a circular table around which are placed 25 chairs, labelled in anticlockwise order A, B, C, etc. as far as Y. Each of the 25 members of the club has one of these chairs, on which he always sits when present at a meeting. Each month in the newsletter, the secretary reports the attendance in the form of a problem for the absent members to solve. Two of these problems are reproduced below.

You, like the absent members, are asked to solve the problems. The time allowed is $1\frac{3}{4}$ hours. It is suggested that you spend not more than one hour on problem 1 before trying problem 2.

2.   PROBLEM 1: January meeting

Each statement below gives a group of chairs and shows how many of them are occupied. For instance, the first statement indicates that of the seven chairs ABCDEFG exactly four are occupied. The other three are vacant.

Statement	Chairs	Number Occupied
1	ABCDEFG	4
2	BCD	1
3	CDEF	2
4	DEFGHIJKLMN	5
5	EFGHIJKLMNOPQRS	8
6	FGHI	2
7	FGHIJKLMNO	6
8	IJ	0
9	IJKLM	3
10	KLMNO	3
11	KLMNOPQR	6
12	MNOPQRSTUVWXY	6
13	OPQRST	4
14	PQRSTUV	3
15	QRSTUVWXYA	4
16	RSTUVW	2
17	UVW	0
18	UVW	1
19	UVWXYABCDEFGH	5
20	YABCDE	3

To make things just a bit more difficult, two incorrect statements have been included. Reconstruct the attendance, expressing your answer as a row of 0's (vacant chairs) and X's (occupied chairs), starting with chair A on the left and ending with chair Y. Identify the two incorrect statements.

PROBLEM 2: February meeting

Six of the members present were asked to state which of the six chairs nearest to their own (three on each side) were occupied. The six replies are given below.

1. Of these six chairs, only the third to my left is occupied.
2. Only the first and third to my left are occupied.
3. Only the third to my left and the first to my right are occupied.
4. Only the second to my left and the first and third to my right are occupied.

5. The second and third to my right are the only ones of the six occupied.

6. The only vacant chairs of the six are the third to my left and the first to my right.

Can you determine from these answers how many members were present? Credit will be given for indications of method of solution if you have not time to complete it.

255. 1975 Entrance Test I

On the whole it seemed a good idea when our Press-cutting Agent went over to computer. When the service was good it was very, very good, but when there were bugs — well this morning for instance we received these two offerings. I would have thrown them into the WPB, but the office boy enjoys this sort of thing, and in fact puzzled out most of it. I should add that we are the well-established firm of Luxwater & Co. We are in the soap and detergent business, and are interested in relevant news items, particularly those which give us some idea of what our less principled rivals are up to, such as Tideless Ltd, and Dreftwood and Flashback.

Read as much as you can of the two press clippings, but do not put in too much time on it — the office boy says that some words are almost impossible.

1. LIDELELL LLD ALLLLLCED LELLELDAL A LEL BILLLGICAL
    LLLDLCL CALLED BALG, LHICH IL CLAILED LL BE CALABLE LF
    LELLLILG LLLLL FLLL LELLALD LKILL. BALG LILL BE LALKELED
    ELLELILELLALLL LLEL LHE LELL 6 LLLLHL IL LHE LELL CLLLLALL
    ALEA.

2. IJ IJ JEJJJJED JHAJ JJ. J. BJCJFAJJ JIJJ AJJJJJCE HIJ
    JEJIGJAJIJJ AJ CHAIJJAJ JF JHE DEJEJGEJJ GIAJJ DJEFJJJJD AJD
    FJAJHBACJ JAJEJ JHIJ JEEJ. HIJ JEJIGJAJIJJ HAJ BEEJ EJJECJED
    EJEJ JIJCE DJEFJJJJD AJD FJAJHBACJ DECJAJED BEFJJE-JAJ
    JJJFIJJ FJJ 1973/4 JF £73. 31.

256. 1975 Entrance Test II

It was a big break for Special Branch when they located the international terrorist regional HQ over a fish and chip shop near Clapham Junction. The raid pulled in a few of the smaller fry, and also, amongst other things, some ciphering materials. These consisted of neat pages of "keys", apparently torn from a pad. Each page was set out in the same way; each contained 5 lines, and each line consisted of four 5-figure groups. Most of the pages had been destroyed, but pages 1-3 were recovered intact. It was found that by studying these pages it was possible to predict, with reasonable confidence, the content of some subsequent pages.

Pages 1-3 are given below.

What is the middle line of page 4 likely to be?

Page 1

29168	53206	78941	73450
98402	50237	61843	15679
54608	59627	91340	17823
76908	53679	18042	12345
51268	37245	98047	01963

Page 2

05437	43925	78462	05319
76519	35164	76480	29518
87132	71265	73240	39816
21543	02365	78203	95614
03691	79425	80417	35610

Page 3

91350	67180	18347	05124
15928	02793	87921	02156
89361	50984	09427	08416
59416	49718	64827	05468
30165	68239	03147	56190

257.   1976 Entrance Test I

We rented a cottage on Sevensey for our summer holiday last year. It is a small Channel island, and consists of the capital – St Margery – and three smaller villages – St Aimée, St Dottie and St Pancras. In the interests of peace and quiet the number of cars on the island is limited to 50. Apparently each of the 4 parishes is given a quota, and each of the Parish Councils allocates its quota amongst its residents.

We have a couple of boys, and they had a game to see who could collect all the car numbers on the island first. You could not tell who had won because there was so much cheating; they ended up with identical lists containing nearly all the numbers and a few bogus ones – 55 in all.

Here is the list:

013	153	274	414	573	821
032	172	293	415	605	884
045	185	312	433	611	891
051	191	325	465	624	961
064	204	331	484	632	999
083	212	344	503	643	
102	223	363	535	675	
115	242	382	541	681	
121	255	395	554	737	
134	261	401	560	751	

I can vouch for 153, 223, 293 and 573, as they were parked more or less outside our cottage in St Aimée. Also for 045 which belonged to our landlord – who lived in St Pancras and was actually Chairman of the Parish Council there.

a.   Which car numbers did the boys invent?

b.   What would you expect the number to be of the car belonging to the Deputy Chairman of the St Margery Parish Council?

c.   There is actually a fifth parish, the adjoining islet of St Swithin, which we did not visit, and which we gathered has its own quota of 3 cars. What would you expect the numbers of these cars to be?

258. 1976 Entrance Test II

These Rototype typewriters have a printwheel instead of the usual type bars. I was given one for Christm?r; Hs hr lnrs tmpcjg.?jc, Uf.r f.q emkb tolkd klt mha,oaz

259. 1977 Entrance Test

Below are extracts from the inscriptions on the Royal tombs of the ancient Nubian kingdom of Up. Each King had a lucky number which regulated the events of his life. For instance, if his lucky number was 4, 4 days elapsed between his presentation to An (the Sun God) as King and his enthronement, and 4 days elapsed between his death and his burial.

King	Day of Presentation	Enthronement	Death	Burial
Upanup I	Houndday	Ramday	Firstday	Snakeday
Philup	Beeday	Snakeday	Orcday	Catday
Philan	Assday	Beeday	Ramday	Houndday
Anlog	Houndday	Catday	Ramday	Orcday
Angram	Beeday	Ibisday	Ibisday	Orcday
Upanup II	Lastday	Lastday	Assday	Assday
Anlist	Assday	Ramday	Snakeday	Assday
Palovan	Catday	Beeday	Dragonday	Snakeday

Reconstruct the Uppian week, which had 11 days.

260.   1980 Entrance Test

I can't say I'm very keen on programmed learning texts – you know, the ones where the pages follow one another in some peculiar order, and every now and then you have a question that directs you to a choice of page depending on which answer you select. Recently I came across one, a mere 20 pages long, which gave me a lot of trouble. At the top of each page was the number of the page or pages immediately preceding (i.e. 'from'), and at the bottom of the page or pages to turn to next (i.e. 'to'). Unfortunately the printer had omitted to give the actual page numbers, so the binder, I suppose in desperation, assembled the pages as far as possible in 'from' order.

Can you number the pages correctly, given the information below?

From	To
1	1
1	1,2
2,3,5	3,5
2,7,10	6,7,9
3	1,4
4	4
4	8
4	10
6,12,14	11,13,14
8	12,14
8	15
8,11	8
9	4
11	8
13,17,19	16,17,18
15	15
15	19
15,20	–
16	15,20
19	18

261. 1981 Entrance Test

The following 35 sequences of letters were extracted from a 9-by-9 array of letters. Each sequence came from a diagonal line, up or down, in either direction (NE-SW, SW-NE, NW-SE, SE-NW), not necessarily beginning or ending at an edge, but always in a straight line.

Reconstruct the array. There is a uniquely convincing solution.

AO	TLS	QFEO	EWHOO	OTURHE
GHM	CSTS	SEPW	HUHLK	STXICI
GRO	IUSC	SIYN	IESNK	ZVOCON
MVR	MBEE	XKNS	OISGR	NUOTYOM
NVD	MSLH	YEXK	RJBTW	TVHNOCO
RUO	OCJH	EOMHG	WHSTS	APRNALSE
SIO	OFIL	EUKLH	ILATBA	PLMBOTSIY

(For example, from the following 5-by-5 array

A	B	C	D	E
F	G	H	I	J
K	L	M	N	O
P	Q	R	S	T
U	V	W	X	Y

the following sequences could have been extracted:

AGMSY, KGC, QMI, EIM, JD, etc).

262.   1984 Entrance Test

All you've got to read is yesterday's newspaper and you've read it from cover to cover twice already. You've done the crossword, too. Finally, in desperation, you see the solution to the previous day's crossword and wonder if you can do anything with it; unfortunately the first half of the 'ACROSS' solutions has been cut away in response to an advertisement on the other side. The following part of the solution remains:

Cafe 17 Cauldron 19 Pile 21 Tights 23 Animus 24 TUC 25 Stubby 26 Kennel

DOWN: 2 Enact 3 Abdominal 4 Press up 5 Credo 6 Mac 7 Cast off 13 Deception 15 Realist 16 Ransack 18 Rusty 20 Louse 22 Hob

Reconstruct the crossword matrix (it is square, there is some symmetry and 'answers' are separated by black squares), and make sensible guesses for the words and acronyms that must fit in the missing 'Across' lights.

263.　1985 Entrance Test

Keyword Alphabets

　　　A keyword alphabet is formed by writing out a word and following it with the unused letters of the alphabet in order. If a word contains repeated letters, the second and subsequent occurrences are omitted. Since 25 = 5 × 5 it is often convenient to have a 25-letter alphabet by regarding I and J as the same letter — in particular this makes it possible to form keyword squares.

　　　Examples:

HYDROPLANEBCFGIJKMQSTUVWXZ

COMITEABDFGHKLNPQRSUVWXYZ

```
B  R  I  C  K
L  E  H  A  M
P  T  O  N  D
F  G  Q  S  U
V  W  X  Y  Z
```

These are based on HYDROPLANE, COMMITTEE and BRICKLEHAMPTON (the longest word we know without repeated letters).

　　　The following string of letters is a keyword alphabet in a simple disguise:

VIYFPELAGSNOBHWTQCKXRUDMZ

Recover the keyword and suggest how the disguise was generated.

264.   1986 Entrance Test

Many antique dealers, in addition to marking their goods for sale with a visible price tag, add also a 'hidden' price which allows an assistant to work out the 'rock-bottom' price below which he is not allowed to drop, should a customer start to haggle. This hidden price is of course somewhat below the visible price, the amount depending both on the dealer's perceived value of the item for sale and what he paid for it.

The hidden prices can be expressed in letters, derived from the corresponding figures by substitution, using a 10-letter code word or phrase with no letter repeated which can be committed to memory by the assistant. For example:

$$\begin{array}{cccccccccc} S & H & O & P & L & I & F & T & E & R \\ 0 & 1 & 2 & 3 & 4 & 5 & 6 & 7 & 8 & 9 \end{array}$$

Using this example, £535 would be written as £IPI.

Last week I was in such a shop. No sooner had I enquired about an item that had caught my eye than the dealer was summoned to the telephone, so I had a few minutes to look at an assorted selection of price tags. When I got home I wrote down those I could remember (reproduced below). From this I managed to work out the dealer's 10−letter codeword, although not without some difficulty as it turned out that I had misread a couple of the hidden prices.

What was the 10-letter codeword?

Item	Visible Price	Hidden Price (As remembered)
Grandfather Clock	£750	£IOH
Gateleg Table	£400	£AAL
China Candlestick	£1.50	£W.OH
Victorian Chamber−Pot and Ewer	£5.50	£S.HL
Set of Dining−Room Chairs	£1000	£FHL
Picture	£10	£F.LL
Armchair	£25	£FW

Bookcase	£250	£WUL
Dresser	£900	£PHL
Antique Mirror	£80	£IH
Cracked Vase	£3	£0.0H

A Right Royal Puzzle

Our maths wizards have been working their magic again. In honour of Her Majesty the Queen's ninetieth birthday, they've dreamt up a right royal puzzle for one's amusement.

It's made up of seven individual challenges, which range in difficulty.

The answer to each challenge is the name of a famous person from history. Once you have worked out the names, convert each into a single letter, using the same process for all of your answers. Unscramble these to find the 7-letter word that royally connects all the answers – and hence solve the overall puzzle.

Good luck. Get your thinking caps (or should that be crowns?) on.

265. A right royal puzzle: question 1

Un-mix and then mix. What do you get?

CLAY KESTREL	JOULE MINCE
CRYSTAL EEL	LEWD GENOME
CUBA SEACOAST	P. E. PREP
CUBIC SEE	REOCCURS WEATHERISES
JAM TOO, CUTIE	D. V. OAK

266. A right royal puzzle: question 2

We thought we had a nice idea for a teaser but clearly had ambitions beyond our station so we had to railroad this one a little to make part 4 work. If you have a well-trained eye you should be able to think of your own clue for part 3, and then you'll be on track to provide the missing answer.

(1) Count of Savoy (1178–1233)

(2) King of England (1284–1327)

(3) ???

(4) Alexander (1743–1827)

(5) King of Scots (1512–1542)

267. A right royal puzzle: question 3

The celebrity pseudonymizer is a way to generate aliases. Given a person's name, think of someone famous with the same surname and use that person's first name, then think of someone famous with the same first name and use their surname. For example DAVID CAMERON might change to RHONA WALLIAMS (after Rhona Cameron and David Walliams), or JENNA COLEMAN might change to DAVID MARONEY (after David Coleman and Jenna Maroney).

A list of award-winners was put through the celebrity pseudonymizer and the results are below:

JOHNS OGOGO

ULYSSES BONNEVILLE

NATHANIEL PARGETTER

MERRILEE CHAUCER

BOONE REDFORD

What was Nathaniel Pargetter's award for?

268. A right royal puzzle: question 4

What comes next?

(a)   HJAFUW, JLCHWY, LNEJYA, NPGLAC

(b)   F, S, Mn, Kr

(c)   The Magician, The High Priestess, The Emperor, Strength

269. A right royal puzzle: question 5

Many major events have happened in the Queen's lifetime. Here are some headlines that could have been seen in the last 90 years – but in encrypted form. Can you work out who is apparently behind the encryption?

DJCJHGI KLHBAJ EGVKJK EFGNK BC OHBLGBC

JPMGHP WBBB GOPBEGLJK – DJNHDJ WB BK CJM ABCD

LHYDWJ IBJ BK GTTNBCLJP QBHKL KJEHJLGHY DJCJHGI NQ VC

KMBLZJHIGCP MBCK LFJ QBHKL JVHNWBKBNC KNCD ENCLJKL

JCDIGCP PJQJGL MJKL DJHSGCY BC MNHIP EVT QBCGI

PHNVDFL ENCLBCVJK BC OHBLGBC'K FNLLJKL KVSSJH

RJGC-EIGVPJ PVWGIBJH BK NWJHLFHNMC GK THJKBPJCL NQ FGBLB

OHJGALFHNVDF BC DJCJLBEK: PNIIY LFJ KFJJT BK EINCJP

TIVLN INKJK BLK KLGLVK GK G TIGCJL

UVJJC JIBZGOJLF EJIJOHGLJK CBCJLBJLF OBHLFPGY

– ABCDEFGHIJKLMN

270.   A right royal puzzle: question 6

This crossword celebrates 2d and 6d. A phrase that links this event
to this organization (11, 7, 2, 4, 3, 6, 10, 3, 6, 7, 5, 4) runs clockwise
around the perimeter, starting at square 1. Solvers are invited to
complete the phrase in the central, unchecked squares.

The clues are not ordinary cryptic clues. Instead, each clue provides
an overall definition of its solution. In addition, it contains an anagram
of its solution plus one extra letter contained somewhere within. These
extra letters, in clue order, identify a character from history.

For example: *You need this licence to imply some of these definitions!*
Gives POETIC, which is anagrammed in the underlined part, with the
extra letter M. (PS: you have been warned!)

Across

9. This does silence. (9)
10. Central heating in one's bed. (7)
11. Gull that half loves water. (7)
12. Voyager to a star in outer space. (9)
13. On most grounds, very tense. (10)
14. Act of simple mindlessness. (6)
15. Yankee with dang high reputation. (6)
17. What one did who goes early to bed. (5)
19. Takes its chances, stings you. (6)
21. Going beyond the present mood. (10)
26. Geodesic made it cross all the way over. (9)
27. London bridge, also heck of a football team. (7)
28. Listen to the ladies and gentlemen. (7)
29. Corsair, pirate, devil's shipmate. (9)

Down

2. Reign of purple blaze, the kind and gracious Queen. (9)
3. Impact is serious, not light. (7)
4. Prodder and tickler, end never comes. (7)
5. Keep using stamina-building foods. (7)
6. Anniversary gets attention you deserve. (6)
7. Raising of new Italian lady. (9)
8. Glorious Last Rites, for example. (7)
16. Connect climes that are wet and windy. (9)
18. Caused when a bad timekeeper so loves his bed. (9)
20. Such as East Edinburgh? (7)
22. Where a manic clout puts me out? (7)
23. Feel, act, tailor one's touch. (7)
24. Much healthier than sugar on one's meal! (7)
25. Where Governor dispenses time. (6)

271. A right royal puzzle: question 7

One of my friends wanted me to work out which is her favourite Shakespeare play. She said I should begin by making a complete list of the titles of all plays attributed to him and cross out all those that didn't appear in the First Folio. I did this and then she said I should cross out all those that started with a particular letter. Then it got complicated – she said I should cross out all the titles that contain any word starting with a letter earlier in the alphabet than the start of her own name. I did that and she looked at my list, and told me that some of those remaining contained a word she didn't like and I should cross all them out too. I asked why she didn't like that word but got the impression she wanted to change the subject.

Next she made me divide my list into 2 columns based on whether the title ended in a vowel or a consonant – and then told me to cross out one of the columns. Finally she said I should divide the ones left into 2 columns based on whether they had an odd or even number of letters – and again cross out one of the columns.

What a palaver. Still, I got there in the end. Truth is, though, that I'm more of a mathematician than a literary type so it all seemed a bit long-winded. Only interesting thing was that at each stage I was left with a square number of plays.

What is her favourite play?

Tiebreaker Puzzles

The competition in the first *GCHQ Puzzle Book* resulted in 7 people coming equal top. To separate them required a fiendish tiebreaker, and this is below. The competitors were given precisely 7 days to complete as much as they could. As the introduction to the tiebreaker said:

'There are 14 questions in all, so you just have to solve 2 a day! However, by reaching this stage you have all shown that you are highly capable puzzle solvers, and hence this tiebreaker needs to be especially challenging to enable us to determine the winner. Hence, please do submit an entry even if you have not been able to answer all the questions.'

The questions are below. The book referred to is *The GCHQ Puzzle Book*.

Five questions relating to the competition so far, as it appears in the book:

272. Tiebreaker: question 1

Consider the penguin message, i.e. the substitution code which appears around the edge of the pages containing the competition announcement. What word precedes ORDER?

273. Tiebreaker: question 2

What mistake in the penguin message would an expert in printing point out – regarding this book?

274. Tiebreaker: question 3

The inside front cover contains a message in code. Which two 9-letter words appear in it?

275. Tiebreaker: question 4

How does the competition announcement provide a hint to the enciphering method used for the message on the inside back cover?

276. Tiebreaker: question 5

On which page of the book is there an even bigger hint to this encipherment?

Nine puzzles

277. Tiebreaker: question 6

Identify these words:

(a) 1L6R, 4N42, P23UM, 267ON, 725ME, 1R3TORY

(b) 1U3Y, A4EI1, 1A3OM, 4IR1Y, 4U2ER, 1A2A3

(c) BA3O, 1CO1, IM25, 8A6N1R, A6L46TE, 6I19EAD

(d) 68MA762, 185G8537, A9936TI87, 276E98RT1, 34RA3SM, 4A6168M4R

(e) 6I2, U8I4, 7I12R, A54545, 276OM31, 414L355255

(f) 2U2, 5A6, 2U3, CO4Y, S2I6Y, 5O1ISH

(g) SP12, 17P4Y, 1N6E3, 54E2ER

(h) B616, F834, 3M78N, A65E3C, F8B1D2, W89MI5

278. **Tiebreaker: question 7**

A to Z

(You just need to say what A to Z stand for, not the letters after the equals signs)

A = N-m, M, N, T, G, T, ...
B = E&C, LN, W, E, CC, PC, ...
C = ER, P, B, NHG, HSK, GR, SK, ...
D = ER, P, B, NHG, HSK, EC, WK, ...
E = G, KU, J, E, D, C, ...
F = S, M-j, K-s, S-s, H-s, N-w, ...
G = S, O-s, G, A-i, A-m, T-s, ...
H = R, K, S, C or J, H, K, ...
I = I, ZY, D-T, T-I, S-F, S-S, ...
J = S, WH, CT, NG, CW, CW, ...
K = K, V, K
L = Z, G, KP, PL, O, RV, ...
M = LL-D, L, CC, IC, MC, B, ...
N = E, BO, C, HC, BC, GG, ...
O = V, DL, WFC, EFC, B-MU, VS-GMU, ...
P = C, O, S, AG, BG, WG, ...
Q = PdC, S, MP, CR, BR, GdN, ...
R = I, S, DV, JvG, P, L, ...
S = LTC, MB, AR, CH, BR, S-A, ...
T = M, H-M, K, N-K, N-N, M-K, ...
U = GL, SC, CdM, TA, CL, SB, ...
V = B, S, V, P, V, GP, ...
W = TZ, M, WC, WH, X, L, ...
X = G, D, ZX, SN, XT, ZES, ...
Y = W, C-n, C-a, H, H, K-m, ...
Z = K, D, O, T, K, K, ...

279. **Tiebreaker: question 8**

Identify the missing numbers. '$x' means 'the answer to question x'. Explain your answers.

(a)   ..., 27, 19, 18, 17, 16, 15, 14, 13, 11, 10, ?

(b)   5, 1, 20, ?, 10, 2, 50, 200

(c)   3, 4, 3, 8, 5,
      ?, 4, 3, 5, 5,
      3, 4, 3, 5

(d)   7, 14, 32, ?, 134, 186, 308, 384, ...

(e)   ..., 7, 12, 8, 16, 9, 13, 10, 15, 11, 12, 23, 6, 1, 2, ?

(f)   1, 5, 6, 2, 23, ?, 27, 9, 3, 10, 34, ...

(g)   14, 33, 19, 24, 5, 29, 24, 43, 19, ?, ...

(h)

$e	?	$g − $f
$g − $f	$f	$a
$a	$b − $f	$a
$c × $f	$d	$d + $f

280. Tiebreaker: question 9

Lettersearch:

```
A M G I S G Z
H S O H I R E
S T I M U H T
A N E T P K A
D L C H S E R
H A P E I M A
E O A T L A M
Q T T A O S M
E E E U N H A
A D B M A L G
```

281. Tiebreaker: question 10

One of the setters wanted to impress the others with a grid he had created. He explained to them that the grid was like an *Only Connect* connecting wall, but that there were 5 different ways to solve it, and that across the 5 possible solutions, every word was in a set of 4 with every other word precisely once. Unfortunately, in his excitement to explain, he knocked over that most unstable of drinking vessels, a partially filled kwiz koffee kup, and one of the words became unreadable. Can you suggest what the missing entry might have been? (There is more than one possible answer.)

BANJUL	Caius	CURT	Darren
FRANCIS	GALAH	GONERIL	györ
HAIFA	hamlet	Hawk	IAGO
Norwich	ORIOLE	penguin	(spillage)

282.   Tiebreaker: question 11

In the puzzle archive we have a clock – a normal clock with the numerals 1 to 12, and also 13 to 24, written around the clockface. However, recently someone – who presumably wants to get revenge on the setters – has taken to sticking words, numbers, or codes over these numerals and moving the hands to show the time indicated by these.

The setters have solved these puzzles, naturally, and have therefore established that the words, numbers or codes in each case form a sequence going around the clockface. However, as soon as we have solved a sequence and removed the stuck-on words, numbers or codes from our clock, another sequence appears the next day. Below are some of these sequences. In each case below, what time showed on the clock?

(a) SEAHORSE, GOOSE, WOLF, CAT, PIG, EAGLE, OWL, MOLE, FINCH, SHEEP, BEAR, FROG

(b) TNF, PLT, PREFF, OTAE, OMWF, SMV, SFWFN, FMURP, NMNF, PFN, FIFWFN, PLFIWF

(c) SYS, RAE, TDE, BSI, RAM, UMW, ATN, GSS, UIN, IRW, OHS, UCH

(d) FORGET, **CHURCH**, UNPACK, SCENIC, BARTON, PATCHY, JUBHAH, STREAM, CLOSER, QUOTES, LIKELY, CHERRY

(e) BRA, HFR, HHMSB, BDDX, LYYM, FFR, ACJGV, PPFNN, SUJG, NPX, OMIIPC, JXTFLH

(f) SFN, GOD, HUUHP, IQFI, WYMJ, LOL, AOVMU, WDMUJ, NAKD, DGR, TNSJPL, DIMEUF

(g) IY, OW, AU, WS, DO, NE, FT, HE, AE, WT, EA, RD

(h) UBCZYYMW, LOXBXJFQ, ZQKNNRS, ZGQTNJD, XIXDHBJUS, IXOATFMU, GCAFMPBW, FZDXRC, YBQSPDCHB, KQOKKJMKE, JOJQCRYTQGJ, KSWQKQVPJX

283. Tiebreaker: question 12

In each case, complete the sequence – there is more than one possible answer.

(a) Ungodly morons, A bonny doorman, Cheeses and rotundity, Draymen and weirdoes, Darkly hot solution, Widow earns forgery, A dandy outsider, ?

(b) Ghana Brutal Kudzus, Bahamas Glitz Gnu, Kazakh Natural Thugs, ?

And one that's made up – so please explain your answer.

(c) Schlep fuzz Mr Goth qi, I write Flatland sea iambi, Hefty white doorman, Widow seducement, ?

284. Tiebreaker: question 13

The setters play a confusing version of poker, which often results in plenty of tears. Here are a few example hands:

High Card: RULED

Pair: TRUMP

Two pair: TIRED

Three of a kind: BIRDS

Straight: STAIR

Flush: SIMON

Full house: DRAMA

Four of a kind: STONE

Straight flush: STARE

In our last game one setter was annoyed to discover that he had only a high card: TEARS

This quickly turned into frustration as he realized he'd missed the fact that he also had a pair: TEARS

His weeping turned joyful when it was pointed out that he could actually make a very strong three of a kind: TEARS

Well, can you imagine his reaction when another setter declared a straight? TEARS

This facade continued as the two setters simultaneously realized that they both had a straight flush! TEARS

A third setter, whose mind had been wandering as he felt his luck had deserted him, drily declared his winning hand. What was it?

285.   Tiebreaker: question 14

Here is a standard crossword, with words separated by black squares, not bars, and connected (i.e. a route can be traced from any letter to any other).

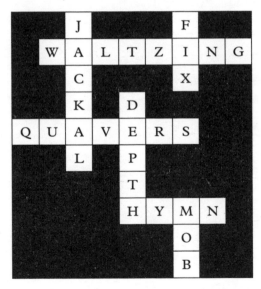

It contains all the letters of the alphabet, and is of size 10 (rows) 9 (columns) = 90. Design the smallest possible connected crossword, in size terms (rows × columns), which contains all the letters of the alphabet. Letters can appear more than once, as in the example above.

Fun with flags

I took a list of the names of every country in the world and ordered it alphabetically. From this alphabetic list I have chosen ten sets of three sequential countries and shown you their flags. You need to identify the countries from their flags.

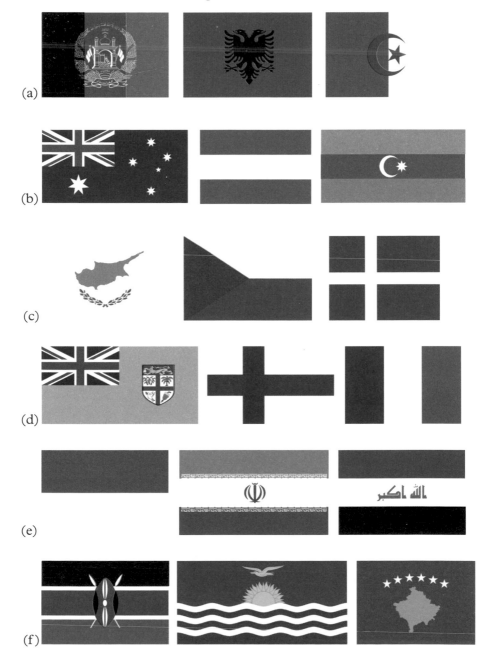

(a)

(b)

(c)

(d)

(e)

(f)

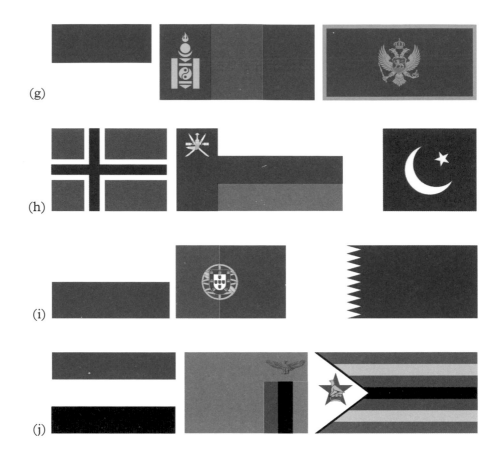

(g)

(h)

(i)

(j)

People VI

Identify these people and divide into pairs:

Picture chains

What are the missing links?

(a)

,?,,?,

(b)

,?,,,?,?,?,

(c)

,?,?,?,,?,

(d)

,?,?,?,

Picture sums V

(a) $\left(\text{} \times \text{😳} \right) - \left(\text{😸} \times \text{😤} \right) - \text{🤐} = ?$

(b)

People VII

Identify the people:

(a)

(b)

(c)

(d)

(e)

(f)

(g)

(h)

(i)

(j)

(k)

(l)

(m)

(n)

(o)

(p)

(q)

(r)

Picture connections II

(a)

(b)

(c)

 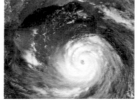 

(d)

People VIII

Identify these people and explain the sequence:

(a)
(b)
(c)
(d)
(e)
(f)
(g)
(h)
(i)
(j)
(k)
(l)
(m)
(n)

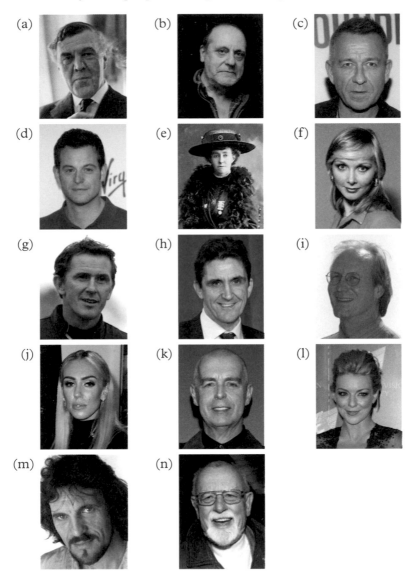

Hints

Many of the puzzles in this book may look more intractable than they really are. In a lot of cases the trick is to approach the puzzle with the right mindset – in other words to think like a GCHQ puzzle setter. So for example when looking at a set of words, don't think of them as words, think of them as a set of letters. And when looking at a set of letters, think whether they could be converted into numbers. And when looking at a set of numbers, think whether they could be converted into letters.

Here are some examples:

What do the members of each set below have in common:

> (1) BOLIVIA, FRAGILE, JACUZZI, PICCOLO, VANUATU
>
> (2) ESTIMATE, ILLINOIS, PITILESS, STRUGGLE

What is the final member of these sequences?

> (3) D, H, L, P, T, ?
>
> (4) 1, 14, 19, 23, 5, ?

In question 1 there isn't really anything Bolivia and a piccolo have in common. But if you look at the words as a set of letters you will see they all have 7 letters. That is something they have in common, but it's not particularly unusual. But if you look more closely you will see that each word ends in a vowel. That's a bit more unusual, and is on the right lines. Having spotted this about the last letters, have a look at the first letters – and you will notice that in each case the first letter of the word comes immediately after the last letter of the word in the alphabet. That is definitely unusual, and now you have the complete answer.

In question 2, again, there isn't really anything Illinois and pitiless have in common. So look closely at the words – they all have 8 letters. But what else have they in common? There is nothing particularly special about the first letters – E, I, P, S – or the last letters E, S, S, E. How about every other letter – ETAE, ILNI, PTLS, SRGL? Again nothing particularly special there. So try looking inside the words. Do they contain hidden words? Well, ESTIMATE contains TIM and MAT, and ILLINOIS contains ILL, LIN and LINO. PITILESS contains several shorter words, but STRUGGLE only really contains RUG. Now you see the connection; the words contain MAT, LINO, TILES and RUG – all things you can put on a floor.

For questions 3 and 4, the letters in 3 don't make a word, and the numbers in 4 don't seem to be a numerical sequence. One thing to try is to convert letters to numbers and vice versa based on where they are in the alphabet. By this method A=1, B=2, C=3, D=4, E=5, F=6, G=7, H=8, I=9, J=10, K=11, L=12, M=13, N=14, O=15, P=16, Q=17, R=18, S=19, T=20, U=21, V=22, W=23, X=24, Y=25 and Z=26.

Using this conversion on the questions turns them into:

      3. 4, 8, 12, 16, 20, ?

and

      4. A, N, S, W, E, ?

Now the answers are clear. Question 3 is every 4th letter in the alphabet, so X is the final member. And question 4 spells out ANSWER, so the final member is 18, representing the R.

This sort of approach will help with many of the questions. Look at first letters, last letters, central letters, every other letter. Look if you can add a letter to the front, or at the end, or in the middle, or if you can remove a letter somewhere. Look for hidden words, or hidden words in reverse. Look for the letters of the alphabet running through a sequence.

When converting letters to numbers and vice versa, there are ways other than A=1, B=2, etc. Letters could be converted into their Scrabble score (see Appendix), or possibly via their positions on a typewriter keyboard or mobile phone keypad (ABC=2, DEF=3, GHI=4, JKL=5, MNO=6, PQRS=7, TUV=8, WXYZ=9).

Letters may need to be interpreted as their NATO Phonetic alphabet equivalents: A=ALPHA, B=BRAVO, etc (see Appendix).

Some number puzzles in the book may be mathematical, but they may also be based on how the numbers are written as words (ONE, TWO, THREE, etc). The British method of writing numbers is used in this book so 108 would be written ONE HUNDRED AND EIGHT. Numbers may also refer to elements in the periodic table, or their symbols (see Appendix).

As well as English we use some French and German, particularly the numbers:

> French: UN, DEUX, TROIS, QUATRE, CINQ, SIX, SEPT, HUIT, NEUF, DIX, ...

> German: EINS, ZWEI, DREI, VIER, FUNF, SECHS, SIEBEN, ACHT, NEUN, ZEHN, ...

> And we also use Roman numerals: I, II, III, IV, V, VI, VII, VIII, IX, X, ...
> (I=1, V=5, X=10, L=50, C=100, D=500, M=1000)

You may also find numbers in other languages on occasion.

There are a number of common themes that run through this book, and having these in mind will help with many of the questions. Some of these – The Periodic Table, US States and US Presidents, NATO Phonetic alphabet, the Greek alphabet, Morse Code, Scrabble values, Braille and Semaphore – are listed in the appendix. We are also fond of some shorter sets:

> Cardinal Points: North, East, South, West (and North West, North East, South West, South East)

> Rainbow: Red, Orange, Yellow, Green, Blue, Indigo, Violet

> Tonic Sol-fa scale: Do, Re, Mi, Fa, So, La, Ti

> The 12 days of Christmas: 12 Drummers drumming, 11 Pipers piping, 10 Lords a-leaping, 9 Ladies dancing, 8 Maids a-milking, 7 Swans a-swimming, 6 Geese a-laying, 5 Gold rings, 4 Calling birds, 3 French hens, 2 Turtle doves and a Partridge in a pear tree

> Nine Muses: Calliope, Clio, Erato, Euterpe, Melpomene, Polyhymnia, Terpsichore, Thalia, Urania

Snow White's Seven Dwarfs: Bashful, Doc, Dopey, Grumpy, Happy, Sleepy, Sneezy

Planets: Mercury, Venus, Earth, Mars, Jupiter, Saturn, Uranus, Neptune (and until recently, Pluto)

Zodiac: Aries, Taurus, Gemini, Cancer, Leo, Virgo, Libra, Scorpio, Sagittarius, Capricorn, Aquarius, Pisces

Wonders of the World: Colossus of Rhodes, Great Pyramid at Giza, Hanging Gardens of Babylon, Lighthouse of Alexandria, Mausoleum at Halicarnassus, Statue of Zeus at Olympia, Temple of Artemis at Ephesus

Mohs scale of hardness: Talc (1), Gypsum (2), Calcite (3), Fluorite (4), Apatite (5), Orthoclase feldspar (6), Quartz (7), Topaz (8), Corundum (9), Diamond (10)

There are also larger sets we like: Olympic host cities, London tube lines, countries and capital cities, books of the Bible, Popes, symphonies and their composers; days of the week, months of the year, units of measurement and animals (especially penguins!) may make an appearance too.

We naturally take an interest in UK geography and the Royal Family. Flags, currencies and vehicle registrations are interesting nationally and internationally.

You may be able to tell something about our literary and musical tastes from the questions. We read Shakespeare, Tolkien, Austen, Poe, Dickens, Lewis Carroll, P.G. Wodehouse, C.S. Lewis, A.A. Milne, J.K. Rowling's Harry Potter books and, of course, Conan Doyle's Sherlock Holmes stories. We grew up with Thomas the Tank Engine, the Wombles, Teenage Mutant Ninja Turtles and the Mr Men. We watch James Bond films and Doctor Who. We listen to ABBA, The Beatles, David Bowie, Bob Dylan and Prince. We read poetry and enjoy musicals.

In our leisure time we play Scrabble, Monopoly and various card games including Pokémon but also Darts and Snooker. So as well as Scrabble scores for letters you will come across Monopoly properties; numbers around a Dartboard; and the values of Snooker balls.

As you would expect from a GCHQ puzzle book there are questions that involve breaking codes. By far the most common is simple substitution cipher

– and this and other ciphers are explained in the How to Solve a Cipher section, which follows this section.

As you would also expect from a GCHQ puzzle book there are some mathematical puzzles – but only recreational maths knowledge should be needed. Some series used are:

> Triangular numbers: 1, 3, 6, 10, 15, 21, ...
>
> Square numbers: 1, 4, 9, 16, 25, 36, ...
>
> Prime numbers: 2, 3, 5, 7, 11, 13, 17, 19, 23, ...
>
> Powers of two: 1, 2, 4, 8, 16, 32, 64, 128, ...
>
> Fibonacci sequence: 1, 1, 2, 3, 5, 8, 13, 21, 34, ... (each number is the sum of the previous 2)
>
> Pi (3.14159265358979...) comes up a few times.

Sometimes different bases are used. Popular ones are:

> Binary (base 2): 1, 10, 11, 100, 101, 110, 111, 1000, 1001, 1010, ...
>
> Hex (base 16): 1, 2, 3, 4, 5, 6, 7, 8, 9, A, B, C, D, E, F, 10, 11, 12, ...

Other bases are sometimes used too.

Finally, there are a number of regular question types for which an explanation may help:

Where: These questions consist of a list of words. Although written in columns for convenience the words are just one list that should be read from left to right and from top to bottom. The question itself asks you where in the list an additional word should be placed.

The way the puzzle is constructed is that the words in the list can be divided into sets depending on words associated with them, and the way these words are associated. Each set is of the same size and has words associated in a different way, and these associated words are in alphabetical order through the list. Typically there are 7 sets of 8 words, or 8 sets of 7 words, though other arrangements may appear. The additional word belongs in one of the sets, and you have to work out where in the list the associated word fits alphabetically, and hence where the word itself should appear in the list.

To make things clearer, here's a small example:

Where does REVEL fit in the following list:

REVILED   PAGE   FIELD   BULLOCK   PART

You might notice that three of these six words can be paired with names to get women who have won the Oscar for Best Actress (Geraldine Page, Sally Field and Sandra Bullock) and the other two words can be paired with words that you get from reversing their spelling (REVILED pairs with DELIVER, PART pairs with TRAP). Furthermore these five paired words are in alphabetical order:

DELIVER   GERALDINE   SALLY   SANDRA   TRAP
REVILED   PAGE   FIELD   BULLOCK   PART

So REVEL must belong to the list of words that can be reversed, and it would be paired with LEVER. Alphabetically LEVER fits between GERALDINE and SALLY, and so the answer is that REVEL fits between PAGE and FIELD.

DELIVER   GERALDINE   *LEVER*   SALLY   SANDRA   TRAP
REVILED   PAGE   *REVEL*   FIELD   BULLOCK   PART

Which: These questions consist of a list of 55 words, which can be divided into 10 sets, all of different sizes. Put another way there is one set of 10 words, one set of 9 words, one set of 8 words and so on down to one set of 2 words and one set of just 1 word. You need to identify the word in the set of one.

The way these questions are constructed is that the words in each set are connected to the size of the set. So the words in the set of 10 have some connection with ten, the words in the set of 9 have some connection with nine, and so on. Again an example might help – in this case cut down to just sets of sizes 1 to 5.

Which word is in the set of one?

ARAMIS	ARCTURUS	ATHOS	BRIAN	CANOPUS
DIVE	GIVE	HITHER	JIVE	LIVE
ON	PORTHOS	PROCYON	RIGEL	SIRIUS

In this case it may be that the names ATHOS, PORTHOS and ARAMIS stand out. These are the THREE Musketeers. You might spot some of the other names are those of stars – ARCTURUS, CANOPUS, PROCYON, RIGEL and SIRIUS. There are five of them, so it's a FIVE STAR solution! Looking at what's left you might spot the four words containing IV, the roman numeral for 4. Of the remaining three words, both HITHER and ON form new words if suffixed by TO (HITHERTO, ONTO), which is a homophone for TWO. So the answer is BRIAN. This is associated with ONE, as a famous BRIAN is BRIAN ENO, and ENO is ONE reversed.

So the 5 sets are:

> 5 – Fivestar: ARCTURUS, CANOPUS, PROCYON, RIGEL, SIRIUS.
>
> 4 – IV: DIVE GIVE JIVE LIVE
>
> 3 – Three Musketeers: ATHOS PORTHOS ARAMIS
>
> 2 – *to: HITHERto, ONto
>
> 1 – ENO: BRIAN

A to Z: In these questions the letters A to Z before the '=' represent single-word members of a particular set of items that start with these letters, and the letters after the '=' represent the initials of people or things that form all or part of that member of the set.

Chains: In these questions a chain can be built up from pairs of words or names that overlap in some way. You have the first and last members of the chain and have to work out the missing links.

Sums: In Sums questions you are presented with a strange-looking sum. Sometimes it consists of words, and sometimes pictures. There is a number associated with each word or picture – within a question these associations will all be of the same kind. You need to identify the numbers and then calculate the result of the sum. This result will also have a word or picture associated with it of the same kind as those in the question and this word or picture is the answer.

For example:

> (evens × ether − eon) / tow =

To solve this you need to spot the connection between all the words, which is that they are all anagrams of numbers. Using the associated numbers gives:

$$(7 \times 3 - 1) / 2 = 10$$

The answer to the question is therefore 'net', an anagram of 'ten'.

How to Solve a Cipher

A cipher is an algorithm that is used along with a secret *key* to encrypt a message, known as the *plaintext*, to produce a seemingly meaningless message, known as the *ciphertext*.

A well-designed cipher is constructed in such a way that even if you have a complete understanding of the encryption algorithm, you still need the key to read the plaintext.

In this section we are going to describe some classical ciphers, explain how to encrypt and decrypt them if you know the key, and then how to break them when you don't! We'll leave a couple of exercises for practice.

Caesar cipher

The Caesar or shift cipher is named after Julius Caesar. To encrypt a message, we take each letter and shift it by a number of places secretly agreed in advance by the sender and receiver. For example, with a shift of 3, A → D, B → E, C → F, and when we go past Z we cycle back to the beginning of the alphabet, so X → A, Y → B, Z → C. To decrypt we just shift the same number of characters backwards.

Encryption example

```
Key:       5
Plain:     I CAME I SAW I CONQUERED
Cipher:    N HFRJ N XFB N HTSVZJWJI
```

How to solve without the key

Cipher: BUMBOFBKZB FP QEB QBXZEBO LC XII QEFKDP

There are only 26 possible keys (25 if we don't include the particularly poor choice of key where you shift by 0 and don't change any character). It doesn't take long to try them all. To speed things up, we can just look at the first few characters:

1	ATLAN	6	VOGVI	11	QJBQD	16	LEWLY	21	GZRGT
2	ZSKZM	7	UNFUH	12	PIAPC	17	KDVKX	22	FYQFS
3	YRJYL	8	TMETG	13	OHZOB	18	JCUJW	23	EXPER
4	XQIXK	9	SLDSF	14	NGYNA	19	IBTIV	24	DWODQ
5	WPHWJ	10	RKCRE	15	MFXMZ	20	HASHU	25	CVNCP

A few of these options look like they could plausibly be beginnings of words, so we can go back and expand anything that looks plausible until either we convince ourselves that it isn't or reveal the plaintext:

1	ATLANEAJYA
6	VOGVIZVETV
7	UNFUHYUDSU
12	PIAPCTPYNP
16	LEWLYPLUJL
20	HASHULHQFH
23	EXPERIENCE IS THE TEACHER OF ALL THINGS

This is an example of a very basic attack known as an *exhaust*, where we just try decrypting with every single possible key, (perhaps saving a little time by having a cheap initial test to throw away anything that doesn't seem plausible without having to decrypt it all). The Caesar cipher is weak because the number of possible keys is very small.

Challenge

VSLBHNERBAGURPRAGRANELGENVYXRRCFBYIVATGURPUNYYRATRF

Keyword cipher

The Caesar cipher is a form of substitution cipher, where the individual letters of a message are replaced by different letters. Another, more complex, form of a substitution cipher is a Keyword cipher, where the letter replacement is done in a less predictable way. We could completely randomly shuffle the letters and our secret key would be the map of what each letter encrypts to. However unless we have a good memory this probably requires us to write down the key somewhere. In some situations this may not have been a great idea (if we were worried about the key being stolen, for example). Therefore keywords were sometimes used so that we can reconstruct the 'random' permutation of letters when we only have to remember one word or short phrase.

Encryption example

Keyword: THEGCHQPUZZLEBOOKII

The key chart is constructed by removing all the repeated letters from the keyword, and then listing all the unused letters in alphabetical order.

Key chart: ABCDEFGHIJKLMNOPQRSTUVWXYZ
             THEGCQPUZLBOKIADFJMNRSVWXY

Plain: LANGUAGE IS THE KEY TO THE HEART OF PEOPLE

Cipher: OTIPRTPC ZM NUC BCX NA NUC UCTJN AQ DCADOC

How to solve without the key

Cipher: VRY PJGSQOGSP O'US ESFOSUSI LP GLHY LP POX
            OGKJPPOEFS QROHAP ESCJNS ENSLDCLPQ

We may first consider attacking this cipher like the Caesar cipher – perhaps using a dictionary to systematically try each word as a keyword. This approach may work with the aid of a computer, but if we want to solve it by hand, reading an entire dictionary may be too much work.

Your eye may have been drawn to O'US. If we see an apostrophe in ciphertext followed by a single character, there is a good chance that the apostrophe is a possessive and the letter following it is an 's'. However this apostrophe is followed by two characters, so it is likely a contraction. Moreover, there is only one letter preceding the apostrophe, so we may think of i'll or i've. Of course i'll has a double letter and 'O'US' doesn't, so i've seems like a sensible first guess. (Another less common contraction to fit the pattern is o'er.) Let's see what happens if we guess that every O, U and S are i, v and e.

```
--- ---e-i-e- i've -e-ieve- -- ---- -- -i- i-----i--e --i--- -e---e --e------
VRY PJGSQOGSP O'US ESFOSUSI LP GLHY LP POX OGKJPPOEFS QROHAP ESCJNS ENSLDCLPQ
```

Because apostrophes give away quite a lot of information, they are often left out of substitution ciphers.

A common early step is to use frequency analysis. In English some letters are more common than others: *etaoin shrdlu* is a nonsense phrase, which lists the 12 most common letters in rough frequency order. Let's count how many times each letter appears in our ciphertext:

A	B	C	D	E	F	G	H	I	J	K	L	M
1	0	2	1	4	2	4	2	1	3	1	5	0

N	O	P	Q	R	S	T	U	V	W	X	Y	Z
2	7	9	3	2	10	0	2	1	0	1	2	0

The most common letters are S, P, O and L. We already think that the S and O represent e and i, so let's guess that the P and L represent t and a.

```
--- t--e-i-et i've -e-ieve- at -a-- at ti- i---tti--e --i--t -e---e --ea--at-
VRY PJGSQOGSP O'US ESFOSUSI LP GLHY LP POX OGKJPPOEFS QROHAP ESCJNS ENSLDCLPQ
```

We may be feeling pretty confident about our guesses because it gives us the word 'at'. Next we may begin to think about what words can fill the gaps. The word '-e-ieve-' stands out as having not many options. Likely candidates without any context are believed, believes, believer, relieved or relieves. However, we suspect that this word follows 'I've', so only believed or relieved make grammatical sense.

```
                b                              b       b     b
--- t--e-i-et i've relieved at -a-- at ti- i---ttirle --i--t re---e r-ea--at-
VRY PJGSQOGSP O'US ESFOSUSI LP GLHY LP POX OGKJPPOEFS QROHAP ESCJNS ENSLDCLPQ
```

However, if we try and do the same thing for t--e-i-et, i---tti[br]le or r-ea--at- we run into trouble. There are no obvious words that fit any of these patterns. This is the most important lesson of cryptanalysis: **constantly question your guesses and don't be afraid to make mistakes.** It is likely that we've made an incorrect guess somewhere. OGKJPPOEFS is an interesting word because it contains a double letter. Common double letters in English are ee, ff, ll, mm, oo, ss and tt. Let's write out our other most likely options and see if anything stands out: `i---ffible`, `i---llible`, `i---mmible`, `i---ooible`, `i---ssible`, `i---ffirle`, `i---llirle`, `i---mmirle`, `i---ooirle`, `i---ssirle`. This approach isn't infallible, but perhaps one likely word now jumps out at you? Perhaps `infallible`? This could be right, although it would mean that we have also mislabeled the a. Another option is `impossible`. Updating our best guess, that the P is not a t, but an s, and placing the m, p and o, we now have:

```
--- some-imes i've believed as ma-- as si- impossible --i--s be-o-e b-ea--as-
VRY PJGSQOGSP O'US ESFOSUSI LP GLHY LP POX OGKJPPOEFS QROHAP ESCJNS ENSLDCLPQ
```

Now we're really getting somewhere and lots of text is beginning to appear. Continuing in this way, we may guess `some-imes` is `sometimes` and `be-o-e` is `before`.

```
--- sometimes i've believed as ma-- as si- impossible t-i--s before brea-fast
VRY PJGSQOGSP O'US ESFOSUSI LP GLHY LP POX OGKJPPOEFS QROHAP ESCJNS ENSLDCLPQ
```

Note that as we uncover more words it becomes much easier to guess the missing words from context:

```
-hy sometimes i've believed as many as six impossible things before breakfast
VRY PJGSQOGSP O'US ESFOSUSI LP GLHY LP POX OGKJPPOEFS QROHAP ESCJNS ENSLDCLPQ
```

And finally:

```
why sometimes i've believed as many as six impossible things before breakfast
```

We can also look at the cipher substitution chart:

```
abcdefghijklmnopqrstuvwxyz    or in full: abcdefghijklmnopqrstuvwxyz
LE.ISCARO.DFGHJK.NPQ.UVXY.                LEWISCAROBDFGHJKMNPQTUVXYZ
```

As we make some recoveries, we may be able to guess the keyword or phrase – in this case LEWIS CARROLL. Note that one of the weaknesses of a keyword cipher is that letters later in the alphabet are quite likely to encrypt to other letters near the end of the alphabet. It is not a random permutation of the alphabet.

Another technique is to guess other common words. According to the Oxford English Corpus, the 25 most common words in English are:

```
1 the   6 a      11 it     16 he    21 this
2 be    7 in     12 for    17 as    22 but
3 to    8 that   13 not    18 you   23 his
4 of    9 have   14 on     19 do    24 by
5 and  10 I      15 with   20 at    25 from
```

For example, if we had a one-letter word we could be almost certain that it was an 'a' or an 'I'.

Yet another tip if you get stuck is to write out the plaintext alphabet with the ciphertext alphabet underneath it. Perhaps you'll be able to make some deductions by guessing the keyword instead?

Challenge

QZS CSXVINE PIN QZLO MLJZSN VLDD NSTSGD QZS HSWQ OQSJ IH QZS MSHQSHGNX QNGLD

Poem code

Another type of cipher is the *transposition cipher*, in which the elements of your plaintext are arranged into a different order. In a transposition cipher the key shows how to do the rearrangement. One example of a transposition cipher is the poem code, where the key is derived from a poem.

Encryption example

To encrypt a message with the poem code, first choose your poem. We're going to use the first verse of 'Cargoes', by John Masefield:

Quinquireme of Nineveh from distant Ophir
Rowing home to haven in sunny Palestine,
With a cargo of ivory,
And apes and peacocks,
Sandalwood, cedarwood, and sweet white wine.

Now let's choose five words from the poem, for example CEDARWOOD, NINEVEH, IVORY, HAVEN and OPHIR.

We're going to use this string to create a permutation: a list of unique numbers in a mixed-up order. We start by looking at all of the As in the string, and numbering them from left to right. We have two As:

C	E	D	A	R	W	O	O	D	N	I	N	E	V	E	H	I	V	O	R	Y	H	A	V	E	N	O	P	H	I	R
			1																			2								

Then look at the Bs... We don't have any Bs in our example, so move on to the Cs instead. We only have one C, so we label that 3. After D, E, F, G, H the key looks like this:

C	E	D	A	R	W	O	O	D	N	I	N	E	V	E	H	I	V	O	R	Y	H	A	V	E	N	O	P	H	I	R
3	6	4	1					5				7		8	10						11	2		9				12		

We continue in the same way: looking for each letter in alphabetical order, and labelling each instance of that letter with the next number up, from left to right.

C	E	D	A	R	W	O	O	D	N	I	N	E	V	E	H	I	V	O	R	Y	H	A	V	E	N	O	P	H	I	R
3	6	4	1	24	30	19	20	5	16	13	17	7	27	8	10	14	28	21	25	31	11	2	29	9	18	22	23	12	15	26

This string of numbers is our key, which we will use to encrypt the following message:

This type of cipher is known as single columnar transposition. A transposition cipher is any cipher where letters are rearranged.

We write our message in a box underneath our key, using STOP to represent a full stop, and pull out columns from the box according to the order specified by the key. The column labelled 1 has SUIE in it, so that is the start of the cipher. The next column labelled 2 is NTH, and the one after that is TCPE (note that columns have two different lengths), and so on.

C	E	D	A	R	W	O	O	D	N	I	N	E	V	E	H	I	V	O	R	Y	H	A	V	E	N	O	P	H	I	R
3	6	4	1	24	30	19	20	5	16	13	17	7	27	8	10	14	28	21	25	31	11	2	29	9	18	22	23	12	15	26
T	H	I	S	T	Y	P	E	O	F	C	I	P	H	E	R	I	S	K	N	O	W	N	A	S	S	I	N	G	L	E
C	O	L	U	M	N	A	R	T	R	A	N	S	P	O	S	I	T	I	O	N	S	T	O	P	A	T	R	A	N	S
P	O	S	I	T	I	O	N	C	I	P	H	E	R	I	S	A	N	Y	C	I	P	H	E	R	W	H	E	R	E	L
E	T	T	E	R	S	A	R	E	R	E	A	R	R	A	N	G	E	D												

The resulting cipher is:

SUIE NTH TCPE ILST OTCE HOOT PSER EOIA SPR RSSN WSP GAR CAPE IIAG LNE FRIR INHA SAW PAOA ERNR KIYD ITH NRE TMTR NOC ESL HPRR STNE AOE YNIS ONI

However, we don't want to make it obvious where the columns begin and end, so we respace into 5-letter groups:

SUIEN THTCP EILST OTCEH OOTPS EREOI ASPRR SSNWS PGARC
APEII AGLNE FRIRI NHASA WPAOA ERNRK IYDIT HNRET MTRNO
CESLH PRRST NEAOE YNISO NI

This gives the ciphertext. The sender also needs to tell the receiver which words from the poem to use. We have used CEDARWOOD, NINEVEH, IVORY, HAVEN and OPHIR – the 24th, 3rd, 18th, 10th and 6th words of our poem, in that order. We can turn these numbers into letters (A=1, B=2, etc.) to give a 5-letter group: XCRJF. This is the key indicator group, which would be prepended to the message before sending.

To decrypt, the receiver has the same poem, and receives the indicator group. From that they can work out which words from the poem have been selected, and in which order, and from this they can reconstruct the same key. Because the key is 31 letters long, and the message is 112 letters long, the receiver can also reconstruct the exact dimensions of the box.

$$112 = (3 \times 31) + 19$$

So the box needs three full rows and one partial row of 19 entries.

The receiver can now start inserting the ciphertext into the box according to the order given by the key until it hits the boundary:

C	E	D	A	R	W	O	O	D	N	I	N	E	V	E	H	I	V	O	R	Y	H	A	V	E	N	O	P	H	I	R
3	6	4	1	24	30	19	20	5	16	13	17	7	27	8	10	14	28	21	25	31	11	2	29	9	18	22	23	12	15	26
T			S																			N								
C			U																			T								
P			I																			H								
E			E																											

And so on until the whole message has been reconstructed.

How to solve without a key

Cipher:

TTSAB EKRAA TPAED SPASU SEENH OEGAS CYRNE ECERY NCSAT
EINID SEOSM HOACT RAHMW FJSME EDEOT TBBTY EEERE RKPNE
ECOFU QJEYO EMCSD TCMNB OHREA RSNNE FCBTT BOAHH ETELE
IONDN TRROR ESUUS UWNNE OEGGI UNCPO AIIFS IARTB PEDIG
IODDL RTISE NNEUE IEEMN DNEEM TO

Not knowing the key used for a columnar transposition means that not only do we not know the order in which to read out the columns, we also don't know how many columns there are, or anything about how long they might be. However, there is an approach we can take to attack such a message, without knowing anything about the shape of the box.

We are aiming to divide the text into columns that 'fit together' in a way that makes nice words, reconstruct the box from there and then recover the message. A good starting point is to look for any Qs in the message, because a Q is almost always followed by a U in English text. Our message has one Q in it, at position 96. We write the cipher around this point in a column.

O
F
U
Q
J
E
Y

Then we look at all the Us in the message: there are seven. For each one, we write the cipher slightly before and after it into a new column. We only have to check six of the Us, because one of them is in the position just before the Q, and so can't be next to it. Here are the six options:

OP	OR	OE	OU	OG	ON
FA	FE	FS	FU	FG	FN
US	US	UU	US	UI	UE
QU	QU	QU	QU	QU	QU
JS	JU	JS	JW	JN	JE
EE	ES	EU	EN	EC	EI
YE	YU	YW	YN	YP	YE

Of these, the second option looks the best, because none of the potential bigrams looks bad. The J in the 5th row causes problems for options 1, 3, 4 and 5, none of which looks good. Option 6 is also a possibility: FN looks bad but could be the beginning of one word and the start of another. (There is another explanation, which is that the word with Q in it appears at the bottom of the box, and therefore the J does not appear below it, but instead at the top of a new column. We will remember this possibility, and come back to it later if neither of these two options gives anything promising.)

Looking at option 2, we have:

```
OR
FE
US
QU
JU
ES
YU
```

The most common word by far that starts with JU is JUST, so we would like there to be an S in the next column at that position. And directly above it (so one place before it in the cipher) we want to have a vowel, to go after our QU.

There are four positions in the cipher that have a vowel followed by an S; here are each of them written into that position:

```
ORD     ORO     ORD     ORL
FES     FEE     FES     FER
USP     USG     USE     UST
QUA     QUA     QUO     QUI
JUS     JUS     JUS     JUS
ESU     ESC     ESM     ESE
YUS     YUY     YUH     YUN
```

Of these, only the second looks wrong, as the trigram USG seems a bit unlikely (although not impossible, again, there could be a word break here). However, the fourth option gives us an intriguing possibility: could the UST above our QUI be the end of the word JUST as well? Are there any more Js in the cipher?

There is one J, and writing the cipher in that position next to what we have gives:

```
WORL
FFER
JUST
SQUI
MJUS
EESE
```

This is starting to look promising. We should extend our box up a bit, (we've already dropped the last row because it wasn't working so well), to see if the ciphertext before each of these columns also fits together to make a nice word.

HERD
MCOD
WORL
FFER
JUST
SQUI
MJUS
EESE

This also looks good. Let's see if we can add any more columns. We still want the JUS in the penultimate row to be JUST, and it looks like the WORL in row 3 should be WORLD. This means that we want places in the cipher where a D is followed by three letters that can be anything and then a T. There are two of those, but of them has already been used in our fourth column. We can add the other one in:

H	E	R	D	E
M	C	O	D	E
W	O	R	L	D
F	F	E	R	E
J	U	S	T	O
S	Q	U	I	T
M	J	U	S	T
E	E	S	E	B

Everything is still looking good. At this stage it is helpful to keep track of the cipher and which letters we have put into your reconstructed box so far. As a working aid, here is the cipher written out with the cipher that is already in the reconstructed box already highlighted.

T	T	S	A	B	E	K	R	A	A	T	P	A	E	D	S	P	A	S	U	S	E	E	N	H	O	E	G	A	S	C	Y	R	N	E
E	C	E	R	Y	N	C	S	A	T	E	I	N	I	D	S	E	O	S	M	H	O	A	C	T	R	A	H	M	W	F	J	S	M	E
E	D	E	O	T	T	B	B	T	Y	E	E	E	R	E	R	K	P	N	E	E	C	O	F	U	Q	J	E	Y	O	E	M	C	S	D
T	C	M	N	B	O	H	R	E	A	R	S	N	N	E	F	C	B	T	T	B	O	A	H	H	E	T	E	L	E	I	O	N	D	N
T	R	R	O	R	E	S	U	U	S	U	W	N	N	E	O	E	G	G	I	U	N	C	P	O	A	I	I	F	S	I	A	R	T	B
P	E	D	I	G	I	O	D	D	L	R	T	I	S	E	N	N	E	U	E	I	E	E	M	N	D	N	E	E	M	T	O			

Here we have a problem — the E at the bottom of the first column is the same position in the cipher as the E at the top of the last column. These columns can't both be right. We will need to keep an eye on this as we progress.

We can now continue in the same way. The most promising string for extending is FFERE, which appears in the middle of the word DIFFERENT. We would also like the words two rows below that to say IS QUITE, which means that on the left we're looking for a place in the cipher for I, then something else, then another I. There are two possibilities, of which one is obviously nicer than the other.

	P	H	E	R	D	E		
	E	M	C	O	D	E		
	D	W	O	R	L	D		
D	I	F	F	E	R	E	N	T
	G	J	U	S	T	O		
	I	S	Q	U	I	T	E	
	O	M	J	U	S	T		
	D	E	E	S	E	B		

We can now see the top two words ought to be CIPHER and POEM CODE. This gives us the pattern IO.D to look for in the cipher, of which there is only one option that's consistent with the columns that we currently have. Filling this in gives us NDWORLD, which is plausibly from the phrase SECOND WORLD WAR, and ROMJUST, which is probably FROM JUST. At this point, it looks clear that the final row is not correct, so we delete it, and this fixes our problem from earlier where we used an E twice.

Now we find the places in the cipher consistent with our guesses again, and fill them in:

			C	I	P	H	E	R	D	E	S	C	
			P	O	E	M	C	O	D	E	U	S	
S	E	C	O	N	D	W	O	R	L	D	W	A	R
			A	D	I	F	F	E	R	E	N	T	
			I	N	G	J	U	S	T	O	N	E	
	.		I	T	I	S	Q	U	I	T	E	I	
			F	R	O	M	J	U	S	T	O	N	

Now the word starting on the top row looks like DESCRIBED or DESCRIPTION, and on the bottom row it looks like the last word is ONE. Once we find the right piece of ciphertext for the next column, we can start to see KEY spelled out in the 5th row, and POEM below that. After we find the only bits of ciphertext that can fit our guesses we have:

		C	I	P	H	E	R	D	E	S	C	R	I	B	E	D
		P	O	E	M	C	O	D	E	U	S	E	D	B	Y	
S	E	C	O	N	D	W	O	R	L	D	W	A	R	S	T	O
		A	D	I	F	F	E	R	E	N	T	K	E	Y	E	
		I	N	G	J	U	S	T	O	N	E	P	O	E	M	
		I	T	I	S	Q	U	I	T	E	I	N	S	E	C	
		F	R	O	M	J	U	S	T	O	N	E	M	E	S	

The bottom two rows look like the starts of INSECURE and MESSAGE respectively. We also have the word STOP in the 3rd row – this is a common way of indicating the end of a sentence in a message, since we are not using punctuation. Once we fill these in, and then the right parts of the ciphertext, we can see another STOP appear in the 5th row, and the start of HERE on the top row.

		C	I	P	H	E	R	D	E	S	C	R	I	B	E	D	H	E	R	E
		P	O	E	M	C	O	D	E	U	S	E	D	B	Y	S	O	E	A	
S	E	C	O	N	D	W	O	R	L	D	W	A	R	S	T	O	P	A	N	A
		A	D	I	F	F	E	R	E	N	T	K	E	Y	E	A	C	H	T	
		I	N	G	J	U	S	T	O	N	E	P	O	E	M	S	T	O	P	
		I	T	I	S	Q	U	I	T	E	I	N	S	E	C	U	R	E	A	
		F	R	O	M	J	U	S	T	O	N	E	M	E	S	S	A	G	E	

At this point we should take a look at our ciphertext, and see what's been used and what's left. It is worth noting that there are some columns that are only going to be six letters high. Looking at the cipher this way also gives us some help with dividing the remainder into columns: it's now much easier to see the possibilities for which columns are going to be six letters and which are going to be seven letters high. For example, at the end of the ciphertext are 18 unused letters: this must be three columns of six letters, so we can put the column boundaries in without knowing where the columns are to be placed yet.

There are only three columns left with seven letters – the first is obvious because it comes at the start of the cipher, and we can place it at the very left hand side of the box because of the S in the third row. The double T in this column is interesting: what if both rows start with the word THE? Sure enough, there are places in the cipher with a corresponding double H and a double E, and those places fit with the letters we already have in the third row.

T	H	E	C	I	P	H	E	R	D	E	S	C	R	I	B	E	D	H	E	R	E
T	H	E	P	O	E	M	C	O	D	E	U	S	E	D	B	Y	S	O	E	A	
S	E	C	O	N	D	W	O	R	L	D	W	A	R	S	T	O	P	A	N	A	
A	T	E	A	D	I	F	F	E	R	E	N	T	K	E	Y	E	A	C	H	T	I
B	E	R	I	N	G	J	U	S	T	O	N	E	P	O	E	M	S	T	O	P	
E	L	Y	I	T	I	S	Q	U	I	T	E	I	N	S	E	C	U	R	E	A	
K	E	N	F	R	O	M	J	U	S	T	O	N	E	M	E	S	S	A	G	E	

The remaining columns are all 6 letters long, so placing them should be more straightforward, but we continue as before. The last line looks like the second half of the word BROKEN, and two rows above that REMEMBERING. Once we have filled those in with the correct ciphertext, we can see the end of the word UNFORTUNATELY appear, so we can fill that in too, and there are so few possible columns left that only one will fit in each place.

										T	H	E	C	I	P	H	E	R	D	E	S	C	R	I	B	E	D	H	E	R	E
E	I	S	B	A	S	E	D	O	N	T	H	E	P	O	E	M	C	O	D	E	U	S	E	D	B	Y	S	O	E	A	
G	E	N	T	S	I	N	T	H	E	S	E	C	O	N	D	W	O	R	L	D	W	A	R	S	T	O	P	A	N	A	
G	E	N	T	C	A	N	C	R	E	A	T	E	A	D	I	F	F	E	R	E	N	T	K	E	Y	E	A	C	H	T	I
I	M	E	B	Y	R	E	M	E	M	B	E	R	I	N	G	J	U	S	T	O	N	E	P	O	E	M	S	T	O	P	
U	N	F	O	R	T	U	N	A	T	E	L	Y	I	T	I	S	Q	U	I	T	E	I	N	S	E	C	U	R	E	A	
N	D	C	A	N	B	E	B	R	O	K	E	N	F	R	O	M	J	U	S	T	O	N	E	M	E	S	S	A	G	E	

At this point we have finished, because the first column also fits the pattern we wanted in the last column, and we've used all our ciphertext. We now rewrite the box so that the start of the message is in the top left corner, and we have reconstructed the box, without knowing any information about its dimensions.

T	H	E	C	I	P	H	E	R	D	E	S	C	R	I	B	E	D	H	E	R	E	I	S	B	A	S	E	D	O	N
T	H	E	P	O	E	M	C	O	D	E	U	S	E	D	B	Y	S	O	E	A	G	E	N	T	S	I	N	T	H	E
S	E	C	O	N	D	W	O	R	L	D	W	A	R	S	T	O	P	A	N	A	G	E	N	T	C	A	N	C	R	E
A	T	E	A	D	I	F	F	E	R	E	N	T	K	E	Y	E	A	C	H	T	I	M	E	B	Y	R	E	M	E	M
B	E	R	I	N	G	J	U	S	T	O	N	E	P	O	E	M	S	T	O	P	U	N	F	O	R	T	U	N	A	T
E	L	Y	I	T	I	S	Q	U	I	T	E	I	N	S	E	C	U	R	E	A	N	D	C	A	N	B	E	B	R	O
K	E	N	F	R	O	M	J	U	S	T	O	N	E	M	E	S	S	A	G	E										

Finally, we can now reconstruct the key, given that we know the order the cipher was sent in:

1 20 6 25 21 27 10 14 22 28 11 23 7 13 8 12 15 3 9 4 2 24 30 18 19 5 26 29 16 17 31

In fact this has also been derived from 'Cargoes', using the words APES, QUINQUIREME, IN, CEDARWOOD, SUNNY.

Challenge

(This isn't a poem code – but it is a single columnar transposition cipher.)

```
RUWDE HMUNH TWNAR UGTOE TRUNC NASEE LRIEE IHMES OACPA
EQOIA NLENT UMOTS SOOMA TRENT HEIHE
```

Clues to the Starter Puzzles

1. Think of the publisher.

2. Begin by using a calendar.

3. It's as easy as Alpha, Bravo, Charlie.

4. Try using a map.

5. It was the best of puzzles, it was the worst of puzzles…

6. If you solve this it will be quite the (multi-)event.

7. Mamma Mia. Here we go again.

8. You might find a clue down the back of the sofa.

9. Actually two's company – three's a red herring.

10. It looks like the Romans have left something in all these towns.

11. You could sing a rainbow to find a group.

12. Underground, overground.

13. Try colouring in the letters.

14. It is the beginning of the end.

15. Think outside the box to find colourful characters.

16. You can 'county' on us for a clue!

17. Think about the ins and outs of the puzzle.

18. It's still as easy as Alpha, Bravo, Charlie.

19.   It's not the beginning or the end.

20.   Don't wh-ine, it'll be f-ine.

21.   To get started observe that Rovers beat United, therefore they must have lost every other game.

22.   The first answer is apprOXimate.

25.   This should be elementary.

26.   Don't forget a rainbow has two ends. Choose the right one.

28.   Actually their ages are FOUR, FIVE, SIX, SEVEN, EIGHT and NINE.

32.   The sequences have more in common than you might think.

34.   You'll have to movie some of the words around to make the pairs!

36.   What's my line?

37.   One Thousand is equal to ten Hundreds.

38.   There's fury in a jury about perjury in Bury.

40.   Question 40 – or question double-top perhaps?

43.   You might want to address these people more formally if you meet them.

44.   LOOK for the clue in the question.

45.   These questions were set by a fun guy.

46.   Star Wars Day is 0.8.

48.   Pick one letter from each answer.

50.   This is a pet question of ours.

54.   The 10th position – or is it the 17th?

56.   You are looking for an 11-letter word (or name) beginning with G and no repeated letters.

58.   Was D. H. Lawrence a fan of dingbats?

60.   For both parts, the 2kg weight is the furthest to the left.

63.   This is as easy as…

66.   The sum of the goal differences must be 0.

67.   Just use your GENERAL knowledge

69.   Are you starting to flag yet?

71.   They contain an A, but needn't have; they contain one vowel, but shouldn't have.

73.   We think this puzzle is only so so.

76.   When I returned to Britain, guess what I QOUND!

82.   Include the 'The' in the nickname.

85.   Yes you can!

87.   We think this puzzle has legs.

90.   If you're struggling, try asking your friend Scot.

92.   Try singing this puzzle.

94.   For parts (a, b) the COVENTRY CAROL brings JOY TO THE WORLD.

95.   Who'd have thought these people had so much in common?

97.   What a super list this is!

101.   If you have no clue, perhaps this will do.

104.   We had a bit of a mix-up this day.

107.   The letters are in order for one-handed Romans.

109.   You'd normally find these words in very different groups, but you can still pick a letter in each case.

111.   If the × is not a ×, then what can it be? And what must 6 and 4 be?

115.   We can't emphasize enough how tiny the mix-up was.

118.	There's no prize for coming FH.
122.	Remember you're a WKMTLE.
124.	What about us?
125.	We'll ECHO the clue in the question: Find an example.
127.	It's as easy as one, zwei, san.
130.	Elephants And Donkeys Grow Big Ears.
136.	K9 may be a companion on this sequence.
137.	You think it's all over?
138.	They said there'll be snow at Christmas. They said there'll be words to group.
142.	This one might come to you very slowly, or 'Hey Presto', very quickly.
144.	We'll give you 90 minutes to solve this one.
146.	Have you tried mixing paints in Europe?
148.	TAR made the 60 and EG made the ATUR.
153.	How's this for a c(l)ue?
158.	You could pluck an answer out of thin air.
168.	A lot of movies are influenced by the United States.
173.	Go with the flow.
177.	To solve this requires a bit of magic – or should I say voodoo?
180.	Stardate.
182.	The show must go on.
186.	Sounds like 'you' need a clue!
188.	These films are getting longer and longer, Miss Moneypenny.
197.	Shant give you a clue. Wont give you a clue.

199. Music by royal appointment.

203. Is that EMBER, or EMBER, or EMBER?

205. Is the word you have heard.

208. Or rather: Where? Pair...

211. It's as easy as 1,2,3...

213. Happy is Yellow.

218. We think finding the answer is on the cards.

226. A premier mixture.

228. This puzzle may seem hard on the outside, but actually it's easy on the outside.

230. The clue is in the question.

233. We'll keep the clues rowling.

235. You can see the answer with your little eye.

238.   N   U

    W E L R

     S   D

242. Have you never played boulder, parchment, clippers?

245. I hope this clue will set you straight.

252. ll th nswrs r th sm.

Answers

Puzzles

1.

Fictional penguins. From Madagascar, Happy Feet, Toy Story 2, Pingu, Linux, and The Wrong Trousers.

2.

Apricot. The ingredients start with abbreviations for months corresponding to the associated number: MARzipan, APRicots, MAYonnaise, JUNiper berries.

3.

The answers are all letters from the NATO phonetic alphabet, given in alphabetical order:

(a) The Grand Budapest **Hotel**

(b) A Passage to **India**

(c) The Treasure of the **Sierra** Madre

(d) **Victor**/Victoria

(e) **Yankee** Doodle Dandy

(f) **Zulu**

a–e all won **Oscars**.

4.

These are UK towns and cities with the endings missing. The words are paired if they share the same missing part: (AN, WEN) DOVER, (OX, WAT) FORD, (EX, PORTS) MOUTH, (BLACK, LIVER) POOL, (NEW, TOR) QUAY, (NOR, SOU) THAMPTON

5.

Twist. This is the second word in the title of a Dickens novel. The others are first words:

Bleak house, Great expectations, Hard times, Little dorrit, oliver Twist

6.

They are all parts of multi-event sports events:

7: Heptathlon: 100m Hurdles, 200m, 800m, High Jump, Javelin, Long Jump, Shot Put

6: Omnium: Elimination Race, Flying Lap, Individual Pursuit, Points Race, Scratch Race, Time Trial

5: Modern Pentathlon: 200m Freestyle, Cross-country Run, Fencing, Pistol Shooting, Show Jumping

4: Female Gymnastics: Balance Beam, Floor, Uneven Bars, Vault

3: Triathlon: 40km Road Cycling, 1500m Freestyle, 10,000m

2: Biathlon: Cross-country Skiing, Rifle Shooting

1: (Ironically!) Triple Jump

7.

Repeated words in ABBA song titles: Ring Ring (2) + Money Money Money (3) = I Do I Do I Do I Do I Do (5). 'Does Your Mother Know' is also an ABBA song.

8.

The Royal Shield of Arms. Arranging British coins produced since 2008 of the corresponding denominations reveals the Royal Shield.

9.

The words form pairs that can be preceded by THREE:

THREE BLIND MICE; THREE DAY EVENT; THREE FRENCH HENS; THREE LEGGED RACE; THREE LINE WHIP; THREE MILE ISLAND; THREE POINT TURN; THREE RING CIRCUS; THREE WISE MONKEYS

10.

£50. The cost is derived from the Roman numerals that appear in the names of the towns:

£1: wIgan – rIpon

£4: st IVes – tIVerton

£5: hoVe – seVenoaks

£50: pooLe – ayLesbury

The last remark refers to Roman roads.

11.

Seven Deadly Sins: Envy, Gluttony, Greed, Lust, Pride, Sloth, Wrath

Colours of the rainbow: Red, Orange, Yellow, Green, Blue, Indigo, Violet

Ancient Wonders: Colossus, Gardens, Lighthouse, Mausoleum, Pyramid, Statue, Temple

Multiples of Seven: Seven, Fourteen, Twenty One, Twenty Eight, Thirty Five, Forty Two, Forty Nine

SI Units: Ampere, Candela, Kelvin, Kilogram, Metre, Mole, Second

Days of the week: Monday, Tuesday, Wednesday, Thursday, Friday, Saturday, Sunday

Continents: Africa, Antarctica, Asia, Australia, Europe, North America, South America

12.

The Wombles. Wellington, Bulgaria, Orinoco, Alderney

13.

(a)   **G**. Red mixed with Blue makes Purple; Red mixed with Yellow makes Orange; Blue mixed with Yellow makes Green.

(b)   **W**. Red light mixed with Blue light makes Magenta; Blue light mixed with Green light makes Cyan; Red light mixed with Blue light and Green light makes White.

(c)   **Y**. In snooker: Yellow + Yellow = Brown; Yellow + Green = Blue; Red + Pink = Black; Green + Brown = Black; Red + Red = Yellow

14.

(a)   **Peace**. It starts with a vegetable; the others end in fruit.
disapPEAR, grAPPLE, PEAce, pondiCHERRY, subLIME

(b)   **Tomato**. It starts with a male animal; the others end in female animals.
crEWE, hooteNANNY, kitcHEN, nightMARE, TOMato

(c)   **Firenze**. It starts with a classical element; the others end in modern elements.
FIREnze, giLEAD, jARGON, mariGOLD, pumperNICKEL

(d)   **Parishioner**. It starts with a capital; the others end in a country.
catwOMAN, deus ex maCHINA, PARIShioner, pyROMANIA, sCUBA

15.

Groudon. The bands had songs called Red, Blue, Yellow, Gold, Silver and Crystal. These are also names of Pokémon games, and the corresponding characters featured on each box. The Kaiser Chiefs have a song called Ruby, of which Groudon is that game's mascot. Note that Coldplay also have a song called X&Y, which are also the names of two more Pokémon games.

16.

Bedford. Clues refer to abbreviations of the counties for which the town is the county town. Trowbridge is in Wilts, Lincoln is in Lincs, Reading is in Berks, Aylesbury is in Bucks and Bedford is in Beds.

17.

The words can be preceded by IN or succeeded by OUT.

inFANCY, inFLUX, inKING, inVEST, inWARD

ABout, COPout, HIDEout, RAGout, WIPEout

18.

The answers spell out ANSWER in NATO phonetics:

(a) **Alpha**

(b) **November**

(c) **Sierra**

(d) **Whiskey**

(e) **Echo**

(f) **Romeo**

19.

e.g. CHIMNEYS, FIRMNESS or CALMNESS. The sequence is formed of 8-letter words, containing AB, CD, EF, GH, IJ, KL and finally MN as the middle two letters.

20.

The pig is in the **PORCh**. The dog is CANine, so is in the CANal. The cat is FELine, so is on the FELl. The sheep is OVine, so is OVer there. The pig is PORCine, so is in the PORCh.

21.

To get started, observe that Rovers beat United, therefore they must have lost every other game. This tells us that Wanderers beat Rovers and drew their other two games, from which we can deduce the scores of all the Wanderers games, then all the Rovers games and, finally, the score between United and City.

Team	Points	Goals For	Goals Against	Goal Difference
Wanderers	5	3	2	1
United	4	8	6	2
City	4	2	5	−3
Rovers	3	3	3	0

United	1	v	3	Rovers
Wanderers	0	v	0	City
City	1	v	0	Rovers
Wanderers	2	v	2	United
City	1	v	5	United
Rovers	0	v	1	Wanderers

22.

(a) apprOXimate	(k) stRATagem
(b) boonDOGgle	(l) interreGNUm
(c) meadOWLark	(m) bEELzebub
(d) bilLIONaire	(n) stacCATo
(e) wildeBEEst	(o) verBATim
(f) cheMOTHerapy	(p) siGNATure
(g) afTERNoon	(q) COWorker
(h) ePIGlottis	(r) miCROWave
(i) lycANThrope	(s) exPANDAble
(j) beneVOLEnt	(t) contRAVENe

23.

The authors hid their surnames in the text:

Dear GCHQ,

On the occasion of your centenary, we your forefathers thought it would be fitting to write. In our day, cryptologic work was prey to all sorts of catastro**phe: lip pes**tilence and other diseases hampered our work; men of lo**w ilk ins**isted that we were in league with the devil. No**w all is** a lot easier and we hope that those who follow **will es**teem both our work and yours. The torch we passed to you no**w heats to ne**ar stellar temperatures. Your colleagues seem to have all our skill in mathematics and language but to judge by your puzzle books an equally deep knowledge of matters such as 70s music acts (the Brothers Gib**b, ABBA, Ge**nesis, for example).

Our best wishes for your future success

Thomas PHELIPPES, John WILKINS, John WALLIS, Edward WILLES, Charles WHEATSTONE and Charles BABBAGE.

24.

24. If the integer part of the number is expressed as words and the letter indicated by the decimal part is taken, TWENTY FOUR is spelled out: EIGH<u>T</u>EEN, T<u>W</u>ENTY SEVEN, SEV<u>E</u>NTY THREE, NI<u>N</u>ETEEN, <u>T</u>WO, THIRT<u>Y</u> THREE, <u>F</u>OURTEEN, F<u>O</u>RTY TWO, ONE H<u>U</u>NDRED AND NINE, THI<u>R</u>TY FOUR.

If the integer part of the number is expressed as the element with that atomic number and the letter indicated by the decimal part is taken, NOT THIS ONE is spelled out: ARGO<u>N</u>, C<u>O</u>BALT, TAN<u>T</u>ALUM, PO<u>T</u>ASSIUM, <u>H</u>ELIUM, ARSEN<u>I</u>C, <u>S</u>ILICON, M<u>O</u>LYBDENUM, MEIT<u>N</u>ERIUM, SEL<u>E</u>NIUM.

25.

82. Each word can have the second letter changed to make an element which has that atomic number: b**O**ron, n**E**on, i**R**on, t**I**n and l**E**ad.

26.

V. The sequence is the first and last letters of rainbow colours: **R**ed, orang**E**, **Y**ellow, gree**N**, **B**lue, indig**O**, **V**iolet.

27.

(4). The calculations use the first, second, third and fourth letters respectively of the numbers, giving:

(1) ONE × TEN = 10

(2) ONE × NINE = 9

(3) FOUR × FIVE = 20

(4) TEN × ELEVEN = 110

28.

Five se**V**en year olds attended. There were four f**IV**e year olds, nine were s**IX**, one was e**I**ght and one was n**I**ne. Each corresponds to the Roman numeral found within the age.

29.

Films with the last letter of the title omitted:

(a) 12 Years a Slav (e)

(b) The Magnificent Seve (n)

(c) Capon (e)

(d) Who Framed Roger Rabbi (t)

(e) Dante's Pea (k)

(f) American Snipe (r)

(g) The Spy Who Loved M (e)

(h) Night of the Demo (n)

(i) Green Car (d)

(j) Johnny Sued (e)

(k) The Man Who Played Go (d)

30.

MVEMJ. The question represents the solar system, with the right-hand side being the first letters of the planets in order of their distance from the sun.

31.

L. The sequence comprises the initial letters of the first names of those in line to the British throne. Order is: Charles, William, George, Charlotte, Louis, Harry, Andrew, Beatrice, Eugenie, Edward, James, Louise, Anne, Peter, Savannah, Isla, Zara, Mia, Lena.

32.

The letters are the first letters of words that form a sequence:

(a) **J** (Months)

(b) **G** (Books of the Bible)

(c) **H** (Chemical elements)

(d) **A** (Greek letters)

(e) **·U** (French numbers)

The second set of questions are the second letters of words from the exact same sequences as the first:

(a) **E** (Months)

(b) **X** (Books of the Bible)

(c) **E** (Chemical elements)

(d) **E** (Greek letters)

(e) **E** (French numbers)

33.

JNFC. All are names of London Underground lines encrypted with the colour as the key, except for one where, conversely, the colour is encrypted with the name as the key: RED/CENTRAL, GREY/ JUBILEE, **PINK/HAMMERSMITH AND CITY**, MAGENTA/ METROPOLITAN, YELLOW/CIRCLE, GREEN/DISTRICT, BLACK/NORTHERN, BLUE/PICCADILLY, BROWN/ BAKERLOO, LIGHT BLUE/VICTORIA.

See 'How to Solve a Cipher' for a fuller explanation of substitution ciphers.

34.

The words concatenate to form the names of films that won a Best Picture Oscar.

CHIC-AGO, BIRD-MAN, BOY-HOOD, GIG-I, HAM-LET, SPOT-LIGHT, O-LIVER, PLATO-ON, PAT-TON

35.

(a) Two, three, ten

(b) January, June, July

(c) Iran, Iraq, Israel

(d) Leopard, lion, lynx

(e) Cabaret, Cats, Chess

36.

The Northern Line. They represent the London Underground stations Mornington Crescent, Charing Cross, Leicester Square and the Oval, which are all stations on the Northern Line.

37.

All the answers contain 'hundred' or '100'.

(a) Hundred Acre Wood

(b) A hundredweight

(c) FTSE 100

(d) The Hundred Years War

(e) One Hundred Years of Solitude

(f) Chiltern Hundreds

(g) The Hundred Days

(h) Hundred to One

(i) Hundred Islands National Park

(j) Hundred Reasons

38.

CENTURY. The sentence indicates PEN → PENURY and US → USURY – thus CENT → CENTURY.

39.

The answer is **Blowin' in the Wind** by Bob Dylan (born Robert Zimmerman):

```
4372 5464 7656   5861 0000 5432 9672 2765
How  many roads must a    man  walk down

0813    9981 1192 4275 0000 5432
Before you  call him  a    man?

4372 5464 7909 5861 0000 9808   2764 7782
How  many seas must a    white dove sail

0813    8074 8244   4506 9041 7811
Before she  sleeps in   the  sand?

9972 0361 4372 5464 9144   5861 9041 1224         3626
Yes, and  how  many times must the  cannonballs fly

0813    9071   3668    0735
Before they're forever banned?

9041 0395   5865 3755   4792 0955   4506 9041 9847
The  answer my   friend is   blowin' in   the  wind

9041 0395   4792 0955   4506 9041 9847
The  answer is   blowin' in   the  wind
```

40.

21. Each number is the sum of the scores for adjacent segments on a dartboard.

41.

(a) 'Alice's Adventures in Wonderland' (which has twelve chapters) resulted from a story that Charles Dodgson told Alice Liddell and her sisters on a trip on the river Isis on 4 July 1862; (b) and (c) 'two short athletes' and 'met a day apart' are anagrams of the titles of two of the twelve chapters, 'Who Stole the Tarts?' and 'A Mad Tea Party' respectively. Within the other ten chapters (d) 'The Queen's Croquet-Ground' opens with white roses (York) being painted red (Lancaster) by two, five and seven; (e) 'Advice from a Caterpillar' ends with Alice being 9 inches high and in (f) 'A Pool of Tears' Alice frightens a mouse by saying '*Où est ma chatte?*'

42.

Evaluating the expression gives 16, the atomic number of **sulphur**.

Interpreting m.n as the nth letter of element m, the numbers spell out **darmstadtium**.

Removing all but the .'s and -'s gives: .-.-.....-...-.-. , which is the Morse code representation of **arsenic**.

43.

e.g. **Georgie**. These are alternative names for British monarchs from Queen Anne to Queen Elizabeth II in order.

44.

All of the words can be followed by 'VISION'.

45.

Varieties of mushroom:

Forest Lamb, Green Pinball, Green Skinhead, Leathery Goblet, Orange Golfball, Pocket Plum, Pod Parachute, Rooting Shank, Rubber Ear, Sulphur Knight, Waxy Crust, Yellow Brain

46.

> **3.1.** If the dates are written as day/month, then Groundhog Day is 2/2 =1, Bastille Day is 14/7=2, Valentine's Day is 14/2=7, and Hallowe'en is 31/10 = 3.1.

47.

> (a) **CHICKEN.** Writing the sequence in a grid, and adding this word, the phrase WHY DID THE CHICKEN CROSS THE ROAD is spelled out.
>
> ```
> I D Y H H T S
> D T D W E O S
> E H A O R R C
> C H I C K E N
> ```
>
> (b) **To get to the other side.** All other answers are also valid!

48.

Centenary. Use the number to index into the word and take that letter.

(a) Capricorn One (C)

(b) George VI (E)

(c) Fun Boy Three (N)

(d) Fantastic Four (T)

(e) Tea for Two (E)

(f) Connect 4 (N)

(g) Earth 2 (A)

(h) Saturn V (R)

(i) July 4 – Independence Day (Y)

49.

WALLY (ODLAW) fits between BLOBBY (MR) and ALLEGRETTO (OTTER).

ME	WARD	BOLD	VESPER	PREMATURE
Answer	*Bard*	*Bashful*	*Bat*	*Burial*
TELEMACHUS	BLACK	STICKS	JUGGLING	GREEDY
Camel	*Cat*	*Chop*	*Club*	*Deer*
PATIENT	PAGODA	SPONGEBOB	ALERT	SONG
Doc	*Dog*	*Doodlebob*	*Dopey*	*Earth*
LIGHTER	CHEERFUL	SAD	SCOTCH	BART
Fighter	*Grumpy*	*Happy*	*Hop*	*Hugo*
KITT	PURLOINED	UNOILED	SIDED	DATA
Karr	*Letter*	*Lion*	*Lop*	*Lore*
WORLD	BONK	RIVER	BLOBBY	**WALLY** ALLEGRETTO
Mad	*Monk*	*Moon*	*Mr*	***Odlaw*** *Otter*
DOGGY	HAUNTED	TOTEM	CORN	OVAL
Paddle	*Palace*	*Pole*	*Pop*	*Portrait*
POETIC	PROTECTION	HANGER	START	CONNECTING
Principle	*Racket*	*Ranger*	*Rat*	*Rod*
VOGUE	LIFTING	AWAKE	BLESSED	STUPID
Rogue	*Shop*	*Sleepy*	*Sneezy*	*Somethin'*
REDIPS	POGO	WATCH	MUCH	SPIN
Spider	*Stick*	*Stop*	*Too*	*Top*
PHOEBE	MARIO	EARLOCK	LIZARD	CONQUEROR
Ursula	*Wario*	*Warlock*	*Wizard*	*Worm*

Themes are:

Second word in a UK number-one-song title: *Answer* ME, *Earth* SONG, *Mad* WORLD, *Moon* RIVER, *Mr* BLOBBY, *Somethin'* STUPID, *Too* MUCH

Initial letter can be changed to get a Dungeons and Dragons (5th edition) class: WARD/*Bard*, LIGHTER/*Fighter*, BONK/*Monk*, HANGER/*Ranger*, VOGUE/*Rogue*, EARLOCK/*Warlock*, LIZARD/*Wizard*

'Antonyms' of the seven dwarfs: BOLD/*Bashful*, PATIENT/*Doc*, ALERT/*Dopey*, CHEERFUL/*Grumpy*, SAD/*Happy*, AWAKE/*Sleepy*, BLESSED/*Sneezy*

Can be followed by a hand-held sporting implement: VESPER *Bat*, JUGGLING *Club*, DOGGY *Paddle*, TOTEM *Pole*, PROTECTION *Racket*, CONNECTING *Rod*, POGO *Stick*

Works of Edgar Allan Poe: (The) PREMATURE *Burial*, (The) BLACK *Cat*, (The) PURLOINED *Letter*, (The) HAUNTED *Palace*, (The) OVAL *Portrait*, (The) POETIC *Principle*, (The) CONQUEROR *Worm*

Contains the reversal of an animal: TE**LEMAC**HUS/*Camel*, G**REED**Y/*Deer*, PA**GOD**A/*Dog*, U**NOIL**ED/*Lion*, ALLEG**RETTO**/*Otter*, S**TAR**T/*Rat*, **REDIPS**/*Spider*

Can be preceded by a word ending in –OP to form a word: *Chop*STICKS, *Hop*SCOTCH, *Lop*SIDED, *Pop*CORN, *Shop*LIFTING, *Stop*WATCH, *Top*SPIN

Evil twins: SPONGEBOB/*Doodlebob*, BART/*Hugo*, KITT/*Karr*, DATA/*Lore*, **WALLY/Odlaw**, PHOEBE/*Ursula*, MARIO/*Wario*

50.

ALLEY, BOB, POLE, THUNDER: all precede CAT.

BULL, HANG, HOT, WATCH: all precede DOG.

51.

Inky. The friends of the Ghosts all have partners: Bonnie & Clyde, Nutsy & Blinky, Pinky & The Brain. Clyde, Blinky and Pinky were 3 of the 4 ghosts in Pac-Man; the 4th was Inky (who was described as bashful).

52.

    (a) South West Coast Path

    (b) Cotswold Way

    (c) Offa's Dyke

    (d) Monarch's Way

    (e) St Cuthbert's Way

53.

The statement is formed from pairs of letters used to represent locations on UK vehicle registration plates. Three different systems have been used since 1962, representing the three different encodings in the question. All three statements read as follows:

ME RR YX MA SA ND AH AP PY NE WY EA RF RO MU SA LL

54.

The sequence of words contains ZA, YB, XC, etc – and there isn't a word containing QJ consecutively.

55.

6 men.

The men who only travel from A to B and back require two rations to do so, so can each supply an additional ration for those travelling further.

The man travelling to P needs four rations to get from B to P and back, which he cannot carry alone, so a second man must also travel from B to C. Between them they will need a total of 6+4 = 10 rations, of which they can carry three each initially, so they require four other men to carry the four additional rations they need.

Thus: six men set out from A to B. Four of them return to A, giving an extra ration to each of the other two men, and leaving two rations at B. The other two men travel to C, each arriving with two rations. One of them leaves one ration at C, heads back to B, picks up a ration there, and heads back to A. The other uses his two rations to travel to P and back, then picks up a ration at C, heads back to B, picks up the last ration at B and heads back to A.

56.

GLASTONBURY. This is another 11-letter word, starting with G, with no repeated letters.

57.

Lobotomised a Mule = Automobile Models. These are all anagrams.

(a) Ford Granada

(b) Vauxhall Senator

(c) Seat Cordoba

(d) Nissan Pathfinder

(e) Mitsubishi Carisma

(f) Toyota Starlet

(g) Citroen Grand Picasso

(h) Fiat Punto

(i) Aston Martin Rapide

(j) Renault Espace

58.

Sons and Lovers. S on S and L over S

59.

DUNCE.

1: (dENOunce) DUNCE

2: (Two's company) ROYAL SHAKESPEARE

3: (Three French Ns) NATIONAL, NOM, NORD

4: (The Sign of Four) GESTURE, HINT, NOTICE, PORTENT

5: (High Five) DRUGGED, EMINENT, LOFTY, RAISED, TALL

6: (AT sixes …) ATheist, ATone, ATop, ATtempt, ATtired, ATtract

7: (… AND sevens) brigAND, confirmAND, elAND, errAND, quicksAND, remAND, ribAND

8: (Boxing weights) BANTAM, CRUISER, FEATHER, FLY, HEAVY, LIGHT, MIDDLE, WELTER

9: (The whole nine YARDS) BARN, CHURCH, COURT, FARM, GRAVE, JUNK, SCHOOL, SHIP, VINE

10: (TEN reversed) caberNET, corNET, dragNET, driftNET, fishNET, hairNET, interNET, moNET, seine NET, sonNET

60.

(a)

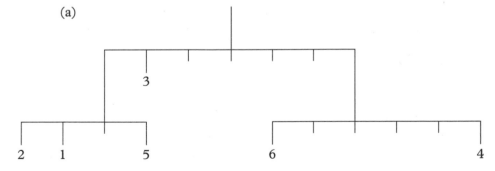

(b)

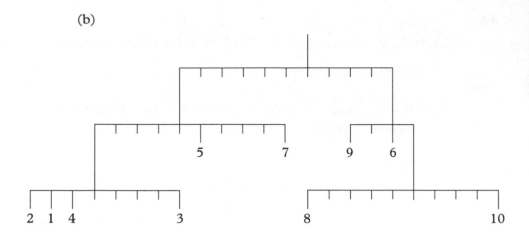

61.

Films with an extra letter inserted. The title is an example of this –
All The Right Mov(i)es (1983).

(a) Hobson's Cho(c)ice (1953) (h) Dun(c)e (1984)

(b) Get Cart(i)er (1971) (i) Sta(i)rgate (1994)

(c) Jaw(a)s (1975) (j) To(r)y Story (1995)

(d) S(i)tar Wars (1977) (k) Fantastic F(l)our (2005)

(e) Tap(a)s (1981) (l) John Tucker Must Die(t) (2006)

(f) Tr(o)on (1982) (m) Kill Lis(z)t (2011)

(g) Beverly Hills Co(o)p (1984)

62.

321. Each number points to the exponents (powers) in its prime
factorization.

$360=2^3×3^2×5^1$

63.

A, B, C, D, to make Ace, Bass, Case, Dace. Four rhyming words.

64.

(a) Writers: Aus**ten** − **Deigh**ton × **Ione**sco = 10 − 8 × 1 = 2 =
Atwood = **A*od**

(b) Planet of the Apes films:

$(5 + 4) / (2 + (n+1) - n) = 9 / 3 = 3 =$ **escape**

The answer is independent of how you start numbering the rebooted films.

65.

AUSTRALIA. The words are anagrams of CHINA, AUSTRIA, SYRIA, LATVIA, GERMANY, ANGOLA, IRELAND, ICELAND, ARMENIA minus the letters: A, U, S, T, R, A, L, I, A.

66.

We begin solving this by filling in the goal difference column. The sum of the goal differences must be zero. This reveals that Rovers didn't score or concede a goal, so we know that all three of their games finished 0-0. Since City finished below Rovers in the group, we also deduce that City can't have won any games. Hence Wanderers cannot have beaten United, since we now know that this would place them above United in the table. Since Wanderers scored 3 against United and only conceded 3 in total, this result must have been 3-3, and the other scorelines can be deduced from this.

Team	Points	Goals For	Goals Against	Goal Difference
United	5	9	7	2
Wanderers	5	4	3	1
Rovers	3	0	0	0
City	1	4	7	−3

Wanderers	1	*v*	0	City
United	0	*v*	0	Rovers
Rovers	0	*v*	0	City
United	3	*v*	3	Wanderers
Rovers	0	*v*	0	Wanderers
City	4	*v*	6	United

67.

Army officer ranks: next in sequence is **FM** = Field Marshal

68.

(a) **Twelve o'clock**: Gin & Tonic, Cheese & Wine, Eric and Ernie, Snakes & Ladders, Chapter & Verse, Accident & Emergency, Stan & Ollie, Arts & Crafts, Mapp & Lucia, Roland & Oliver, Noughts & Crosses, Lock & Key

(b) **Twenty five to twelve**:

```
TACKLE
SWITCH
ENERGY
UNCLIP
NATIVE
MYSELF
FIXING
BISQUE
JOVIAL
GATEAU
HOTPOT
ZODIAC
TACKLE
SWITCH
ENERGY
UNCLIP
NATIVE
MYSELF
```

(c) **Ten seconds before nine**. Replace the numbers by the respective element symbol: Te, N, Se, Co, Nd, S, Be, F, O, Re, Ni, Ne.

(d) Twenty to nine. Replace each country by its capital and take the last letter: RABAT, WARSAW, ROME, YEREVAN, MUSCAT, CONAKRY, BUDAPEST, TOKYO, TALLINN, NAIROBI, LISBON, SKOPJE

(e) **Five past five**. The last letters of the words are TAKE THE FIFTH, and starting at the fifth word, and cyclically stepping through each fifth word from there, the initial letters spell FIVE PAST FIVE.

(f) **Noon**. Using A=1, B=2, etc, subtract the numbers from ciphertext.

```
  CSCNOTXMUTIKICWXNOYAFYQWGBGDOZKWHFQVIXDHKQIIYTSXKXY
- ONETWOTHREEFOURFIVESIXSEVENEIGHTNINETENELEVENTWELVE
= NEXTREDECODETHERESTHWAXRKWSYFSCCTWCQOSPCYLMDKZVSYBT
```

This gives NEXT REDECODE THE REST, so, doing that:

```
  HWAXRKWSYFSCCTWCQOSPCYLMDKZVSYBT
- SIXSEVENEIGHTNINETENELEVENTWELVE
= ONCEMORETWLUIFNOLUNBXMGQYWFYNMFO
```

This gives ONCE MORE, so do it once more:

```
  TWLUIFNOLUNBXMGQYWFYNMFO
- EIGHTNINETENELEVENTWELVE
= ONEMOREAGAINSABUTILBIAJJ
```

This gives ONE MORE AGAIN, so do it one more time:

```
  ELEVENTWELVE
- SABUTILBIAJJ
= NOWYOUREDONE
```

This gives NOW YOU'RE DONE.

The initial letters of the four messages spell NOON.

69.

By rotating their flags through 90 degrees.

70.

(a) The **A**lchemist – Paolo Coelho

The alchemist picked up a book that someone in the caravan had bought. Leafing through the pages, he found a story about Narcissus.

(b) The **B**each – Alex Garland

Vietnam, me love you long time. All day, all night, me love you long time.

(c) The **C**astle – Franz Kafka

It was late evening when K arrived. The village lay deep in snow.

(d) La **D**isparition – Georges Perec

Anton Voyl n'arrivait pas à dormir. Il alluma. Son Jaz marquait minuit vingt.

(e) The **E**xorcist – William Peter Blatty

The blaze of sun wrung pops of sweat from the old man's brow, yet he cupped his hands around the glass of hot sweet tea as if to warm them.

(f) The **F**ountainhead – Ayn Rand

Howard Roark laughed. He stood naked at the edge of a cliff. The lake lay far below him.

(g) The **G**ruffalo – Julia Donaldson and Axel Scheffler

A mouse took a stroll through the deep dark wood. A fox saw the mouse and the mouse looked good.

(h) The **H**obbit – J. R. R. Tolkien

In a hole in the ground there lived a hobbit.

71.

    (a) The A can be replaced by any other vowel and still form a legitimate English word.

    (b) These are not words, but the vowel can be replaced by any other vowel to form a legitimate English word, e.g. FEND, FIND, FOND and FUND.

72.

Each missing word is almost an anagram of ENIGMA, but each is missing one letter.

    (a) Thor's son is MAGNI. (E)

    (b) The comic book company is IMAGE. (N)

    (c) Drake's sister is MEGAN. (I)

    (d) The pine tree state is MAINE. (G)

    (e) The Rolling Stones single is ANGIE. (M)

    (f) The Latin element is IGNEM. (A)

73.

Each is an example of a word consisting of two repeated sections: dodo, tsetse, murmur, Gigi, couscous and hotshots. The title mentions Mum and Dad. Mama and Papa also have this property.

74.

'Here Comes The Night'/Them (track 2); 'I Can't Explain'/The Who (track 6); 'Sorrow'/The Merseys (track 8); 'Shapes Of Things'/ The Yardbirds (track 10); 'Where Have All The Good Times Gone'/ The Kinks (track 12) were all covered by David Bowie on Pin Ups (originally 12 tracks, later 14), the follow-up album to Aladdin Sane.

75.

Krypton, Bismuth.

The full instructions were:

(1) Begin with a list of the 118 elements. Eliminate the 83 which END IN M.

(2) Eliminate the 4 elements which begin with the letter A.

(3) Eliminate the 3 elements which end with the letter R.

(4) Eliminate all the elements which begin with the letter C.

(5) Eliminate all the elements which begin with the letter N.

(6) The remaining elements can be divided into 2 sets of equal size based on whether they are of odd or even length. Eliminate the half which are of EVEN LENGTH.

(7) Eliminate the 3 elements which are of length 5.

(8) Eliminate the 5 elements whose names do not contain the letter T.

(9) Eliminate the element of length 3.

(10) Eliminate the element whose final letter has a Scrabble value of greater than 2.

76.

A **SEAL**. In each case we can change the first letter of the word to the one preceding in the alphabet to obtain the currency of the corresponding country. South Africa has the RAND, Japan has the YEN and Brazil has the REAL.

77.

(a) 1152, 1215, 1521

(b) 1419, 1491, 1941

(c) 1052, 1520, 2015

(d) 1798, 1897, 1987

(e) 1659, 1956, 1965

(f) 1697, 1769, 1967

(g) 1506, 1605, 1650

(h) 1819, 1891, 1981

(i) 1739, 1793, 1973

(j) 1174, 1471, 1714

78.

Authors known by initials, with the first initial in alphabetical order.

(a) Possession by **A** S Byatt: The book was thick and black and covered with dust.

(b) The Voyage of the Dawn Treader by **C** S Lewis: There was a boy called Eustace Clarence Scrubb, and he almost deserved it.

(c) Vernon God Little by **D** B C Pierre: It's hot as hell in Martirio, but the papers on the porch are icy with the news.

(d) A Room with a View by **E** M Forster: 'The Signora had no business to do it,' said Miss Bartlett, 'no business at all.'

(e) The Great Gatsby by **F** Scott Fitzgerald: In my younger and more vulnerable years my father gave me some advice that I've been turning over in my mind ever since.

(f) The Casual Vacancy by **J** K Rowling: Barry Fairbrother did not want to go out to dinner.

(g) The Wonderful Wizard of Oz by **L** Frank Baum: Dorothy lived in the midst of the great Kansas prairies, with Uncle Henry, who was a farmer, and Aunt Em, who was the farmer's wife.

(h) Carry On, Jeeves by **P** G Wodehouse: Now, touching this business of old Jeeves – my man, you know – how do we stand?

(i)   Before I Go to Sleep by **S** J Watson: The bedroom is strange.

(j)   The Sword in the Stone by **T** H White: On Mondays, Wednesdays and Fridays it was Court Hand and Summulae Logicales, while the rest of the week it was the Organon, Repetition and Astrology.

(k)   A Bend in the River by **V** S Naipaul: The world is what it is; men who are nothing, who allow themselves to become nothing, have no place in it.

(l)   Of Human Bondage by **W** Somerset Maugham: The day broke grey and dull.

79.

(a)   The Lord of the Rings: The Fellowship of the Ring

(b)   The Seven Year Itch

(c)   The Artist

(d)   Chariots of Fire

(e)   The Searchers

(f)   Reservoir Dogs

(g)   Star Wars

(h)   The Exorcist

(i)   There's Something About Mary

(j)   Mamma Mia!

(k)   Psycho

(l)   Spider-Man

(m) Spirited Away

(n)   The Man with the Golden Gun

80.

(I AM) OHM. As the riddle suggests, if you remove the symbols (punctuation) and give this one a whirl (rotate it), the famous formula (i.e. Power/Voltage = Current) W/V = I becomes I AM, and WHO becomes OHM. The question references the scientist George Ohm.

81.

Capital cities can be formed from the words and the symbols for the elements:

AGUE – PRASEODYMIUM (PRAGUE), BERN – LITHIUM (BERLIN), HARE – RADIUM (HARARE), LION – ANTIMONY (LISBON), LOON – NEODYMIUM (LONDON), MAMA – SODIUM (MANAMA), MINK – SULPHUR (MINSK), MONO – ACTINIUM (MONACO), QUIT – OXYGEN (QUITO), SOFA – IODINE (SOFIA)

82.

Derby County has the nickname THE RAMS. The first 5 letters are an anagram of EARTH and the last 4 are an anagram of MARS.

83.

(a)   Mohs scale of hardness:

$$\frac{\text{diamond} + \text{feldspar}}{\text{gypsum}} - \text{calcite} = \frac{10 + 6}{2} - 3 = 5 = \text{apatite} = \textbf{ate pita}$$

(b)   Symphonies by Haydn: $6 + 83 + (43 - 8) / 7 = 94 = $ **The Surprise**

(c)   The 'Lorien Legacies' series of Young Adult books by Pittacus Lore:

Book 1: I am Number Four

Book 2: The Power of Six

Book 3: The Rise of Nine

Book 4: The Fall of Five

Book 5: The Revenge of Seven

Book 6: The Fate of Ten

Using numbers from book titles: $(9 - 7)^{\sqrt{(10 - 6)}} = 4 = $ **Number** (or I am Number)

Using book number in series: $(3 - 5)^{\sqrt{(6 - 2)}} = 4 = $ **Fall**

236 GCHQ PUZZLE BOOK II

(d) Addresses: $(30 / 4) \times (52 - 740 / (32 - 12)) = 112\frac{1}{2} =$ **Cheers**

In Bread the Boswells live at 30 Kelsall Street; in the Harry Potter books the Dursleys live at 4 Privet Drive; Mr Benn lives at 52 Festive Road; in The Simpsons 740 Evergreen Terrace has various owners including Mr and Mrs Winfield, Ruth and Laura Powers, and Ned Flanders and family; Paddington lives at 32 Windsor Gardens; in the Harry Potter books Sirius Black lives at 12 Grimmauld Place; Cheers bar is at 112½ Beacon Street

84.

3. The diagram represents a section of a computer keyboard.

85.

Yes. Yes and Oui are cyclic shifts of each other.

86.

These are all anagrams.

Pink Zoom Bearer = Man Booker Prize.

Booker prize-winning novels and authors:

(a) The Luminaries	(f) Hotel du Lac
(b) Heat and Dust	(g) Hilary Mantel
(c) Ian McEwan	(h) Iris Murdoch
(d) Peter Carey	(i) Schindler's Ark
(e) Ben Okri	(j) Salman Rushdie

87.

e.g. **Spider.** These are creatures with 0, 2, 4, 6, 8 and 10 legs respectively.

88.

(a) **Josephine**

(b) **Colin**

(c) **Sue Perkins**

In each round (except the final) the (unique) Baker with <u>no</u> letters of his/her first name in common with precisely one of the three dishes is eliminated. In each round, the (unique) Baker with <u>all</u> letters of his/her first name in common with precisely one of the three dishes is the Star Baker. And in order from bottom to top the initial letters of the remaining dish spell the name of the presenter in reverse.

Round 1: Kenneth eliminated (focaccia), Anita star (Italian grissini). Soda bread.

Round 2: Danny eliminated (biscotti), Tom star (caramel shortbreads). Nice biscuits.

Round 3: Tom eliminated (flaugnardes), Cheryl star (Charlotte royale). Individual suet puddings.

Round 4: Matt eliminated (blue-cheese quiche), Anita star (tarte tatin). Kiwi fruit tart.

Round 5: George eliminated (spicy lamb pasty), Colin star (double-crusted redcurrant pie). Raised game pie.

Round 6: Richard eliminated (lemon soufflé), Ginger star (syrup sponge pudding). Eggnog pudding.

Round 7: Ginger eliminated (low-fat choux puffs), Stanley star (vegan sherry trifle). Plaited wheat-free loaf.

Round 8: Anita eliminated (espresso mousse), Colin star (vanilla macarons). Éclair au chocolat.

Round 9: Josephine eliminated (Cuba rum baba), Cheryl star (chocolate layer cake). Upside-down cake.

Final: Colin star (hazelnut dacquoise). Spanische windtorte, Sfogliatelle.

89.

The four solution words in the top and bottom rows, GOOROO, ZONDA, LIVID and NUANCE, encrypt to form FLECHE, VASTY, GLOBY and SARSAS – each of which is also a dictionary word. This allows the recovery of the initial settings for the ENIG machine, which has offsets of 5 and 7. Continue by entering in sequence the extra letters from the clues – JAKCFKORXFQHVGRCWE // EXXQWJENSLQVVLDSQOTPVW. Since encryption on an ENIG machine is the same as decryption, this recovers the secret instruction DECRYPT A TWENTYFOUR-CELL RECTANGLE IN THE GRID. Continuing with the ENIG settings as they now are, it is possible to decrypt the 3 by 8 rectangle having SCRAIGHS as its top row. The final decrypted answer is REJEWSKI, ZYGALSKI I ('and') ROZYCKI, the names of the Polish team responsible for the initial decrypt of the German Enigma.

1 A	2 F	3 L	4 E	5 C	6 H	E	7 V	8 A	9 S	10 T	Y
11 S	E	A	R	O	A	D	12 A	R	E	A	13 S
14 S	U	S	A	N	I	15 P	L	E	A	S	E
O	16 B	T	17 S	C	R	A	I	G	H	S	A
18 C	A	19 S	U	A	L	20 V	D	21 R	O	22 S	S
I	23 C	O	R	V	E	E	24 S	25 O	U	C	H
26 A	C	R	E	27 E	S	S	O	Y	N	E	A
28 T	A	R	S	29 P	S	30 I	B	A	D	A	N
I	31 P	E	32 N	O	N	C	E	L	33 V	T	T
34 N	O	N	O	E	S	U	35 R	E	I	36 K	I
37 G	O	T	O	S	38 B	L	E	T	S	O	E
39 G	L	O	B	Y	40 S	A	R	S	A	S	S

90.

New and Found. All the words can precede LAND:

Finland, Greenland, Iceland, Jutland, Lapland, Maryland, Rutland – Newfoundland

91. The answers to the 6 questions are:

Q4. All the answers begin or end with **Emma**:

(a) Emmaus   (b) Emmanuel   (c) Emmanuelle

(d) Lemma   (e) Dilemma   (f) ack emma (am)
or pip emma (pm)

Q10. All the groups can be made into words by adding one each of A, E, I, O, U. The pairs are formed from the order the vowels are added:

(a)

bIvOUAckEd	dIscOUrAgE
dUOdEcImAl	UnpOEtIcAl
flUOrInAtE	UnsOcIAblE
fAvOUrItEs	tAmbOUrInE
EUkAryOtIc	rhEUmAtOId
UlcErAtIOn	nUmErAtIOn
mEndAcIOUs	prEcArIOUs
pOpUlArIsE	OUtpAtIEnt
prEcAUtIOn	ExhAUstIOn

(b) The one left over is PRSSN, which represents **Persuasion**.

Q13. Each answer starts with its clue letter. Each 5-letter answer can be made into a 4-letter answer by removing a letter and anagramming. Pairs are:

(a)	ANKLE	(l)	LANE	K
(b)	BRAVE	(v)	VERB	A
(c)	CHALK	(h)	HACK	L

(d)	DIJON	(j)	JOIN	D
(f)	FROZE	(z)	ZERO	F
(i)	ISSUE	(u)	USES	I
(m)	MEANT	(n)	NEAT	M
(p)	PIQUE	(q)	QUIP	E
(r)	REPAY	(y)	YEAR	P
(s)	STINK	(k)	KNIT	S
(t)	TEACH	(e)	ETCH	A
(w)	WRONG	(g)	GOWN	R
(x)	XENON	(o)	OXEN	N

What's left are the removed letters, and what you can make of them is **Mansfield Park**.

Q15. Northanger Abbey. The answers to the clues are formed from the full 100-tile Scrabble set. The first letters in reverse spell FIFTH JANE AUSTEN NOVEL.

(a)	LIMPS	(k)	**ANGER**
(b)	EQUAL	(l)	EVERY
(c)	VOWEL	(m)	NOMAD
(d)	OZONE	(n)	**ABBEY**
(e)	NAIAD	(o)	JEWEL
(f)	**NORTH**	(p)	HOIST
(g)	EqUIP	(q)	TAROT
(h)	TACIT	(r)	FAKIR
(i)	SqUID	(s)	ICING
(j)	URGES	(t)	FOXED

Q17. Pride and Prejudice.

The words follow AND in titles of the form X and Y. The initial letters of the words before AND spell out PRIDE:

Pomp and Circumstance

Romeo and Juliet

Itchy and Scratchy

Decline and Fall

Ebony and Ivory

Q19. Sense and Sensibility. The year 2017 was the 200th anniversary of Jane Austen's death. Austen wrote 6 major novels, and each one provides the answer or theme to a question in this kwiz. In each case the number of letters in the title of the book corresponds to the question:

Q4. **Emma**

Q10. **Persuasion**

Q13. **Mansfield Park**

Q15. **Northanger Abbey**

Q17. **Pride and Prejudice**

Q19. **Sense and Sensibility**

92.

The names are those of the members of familial pop groups.

JACKIE, JERMAINE, MARLON, MICHAEL, RANDY, TITO – The Jacksons

ANNE, BERNIE, COLEEN, LINDA, MAUREEN – The Nolans

ANDREA, CAROLINE, JIM, SHARON – The Corrs

ISAAC, TAYLOR, ZAC – Hanson

KAREN, RICHARD – The Carpenters

MADONNA – Madonna

93.

EROTIC. The links to the numbers 1–10 are in reverse order.

1 (10) EROTIC. Can be preceded by X (Roman 10) to form a word.

2 (9) CHIN, SLAIN. Anagrams of (Nine) INCH NAILS.

3 (8) DAYS A WEEK. Title of Beatles song.

4 (7) ANTISEPTIC, INTERSEPTAL, SEPTEMBER, TRANSEPT. All contain SEPT (French for 7).

5 (6) NAP, POLE, OLE, LEO, EONS (subsets of the Six NAPOLEONS).

6 (5) ABANDON, ABATTOIR, LARGE, NOTABLE, REMOVE, TOWERING. Synonyms for words preceding Five: Maroon, Slaughterhouse, Big, Famous, Take, High.

7 (4) QUEEN, ULSTER, AUTUMN, RAPHAEL, THYMINE, EAST, TEMPERANCE. Members of quartets – Honour cards, Irish regions, Seasons, Ninja turtles, DNA nucleobases, Compass points, Cardinal virtues. Their initial letters spell QUARTET.

8 (3) BLIND MICE, CARD TRICK, LITTLE PIGS, WISE MEN (pairs of words preceded by THREE).

9 (2) ADAMS, ALDRIN, LINUS, PARIS, PENNSYLVANIA, MONACO, HELIUM, CANOPUS, VESTA. All are seconds: US President, man on the moon, Pope (People); Olympic venue, US state, smallest country (Places); element, brightest star, biggest asteroid (Scientific).

10 (1) MY TRUE LOVE GAVE TO ME A PART RIDGE I NA PE ART REE. Spells out what happened on the first day of Christmas.

94.

Letters are taken alternately from two Christmas songs.

(a, b)   JOYTOTHEWORLD
         COVENTRYCAROL

(c, d)   FROSTYTHESNOWMAN
         LITTLEDRUMMERBOY

(e, f)   THEHOLLYANDTHEIVY
         GOODKINGWENCESLAS

(g, h) DINGDONGMERRILYONHIGH
        GOTELLITONTHEMOUNTAIN

(i, j) PLEASECOMEHOMEFORCHRISTMAS
        ISAWMOMMYKISSINGSANTACLAUS

95.

JOHN KERRY. The *s represent the letters that change between the answers:

MARY BEaRd

mArY bERRY

kATy pERRY

matt TERRY

JOHN tERRY

JOHN KERRY

96.

(a) Porthos, Aramis, Athos (The Three Musketeers)

(b) Matthew, Mark, John, Luke (Evangelists)

(c) Daphne, Scooby-Doo, Velma, Shaggy, Fred (Mystery Inc.)

(d) Catherine, Jane, Kathryn, Katherine, Anne, Anne (Wives of Henry VIII)

(e) Bashful, Happy, Doc, Dopey, Sleepy, Sneezy, Grumpy (seven dwarfs)

(f) Kanga, Rabbit, Eeyore, Piglet, Tigger, Pooh, Roo, Owl (Friends of Christopher Robin)

(g) Gandalf, Sam, Legolas, Merry, Gimli, Pippin, Boromir, Aragorn, Frodo (The Fellowship of the Ring)

97.

> **Nation.** The other words can follow Super Mario in the titles to video games, whereas supermarionation is a style of puppetry popularized by Thunderbirds.

98.

(a) Mobile phone text equivalents:

$$\sqrt{\frac{\text{one} + \text{forty} - \text{nine}}{\text{eight} - \text{six}}}$$

The dash between forty and nine is either a hyphen or a minus sign, so the result could be:

$\sqrt{((1 + 40 - 9) / (8 - 6))} = \sqrt{16} = 4 = \text{four} = \textbf{dour}$

or

$\sqrt{((1 + 49) / (8 - 6))} = \sqrt{25} = 5 = \text{five} = \textbf{dive}$

(b) TV programmes containing numbers:

$$\frac{\sqrt{90210 - (7 \times 30)}}{(\text{Seventh Prime})^2 - 13^2} = \sqrt{90000} / (17^2 - 13^2) = 300 / 120 = 2\frac{1}{2} = \textbf{Men}$$

(c) Incidental numbers in song lyrics:

40000 men and women every day ('Don't Fear the Reaper' by Blue Oyster Cult)

Caught in the middle of **105** ('Moonlight Shadow' by Mike Oldfield)

21 years when I wrote this song ('Leaves That are Green' by Simon & Garfunkel)

2 paths you can go by ('Stairway to Heaven' by Led Zeppelin)

$(40000 / 105)$ $(21 / 2) = 4000 = \textbf{Holes in Blackburn, Lancashire}$ ('A Day in the Life' by The Beatles)

99.

(a) **100010.** The sequence is words represented by the Scrabble score of their letters. The whole set is used, so Q, Z and the two blanks are missing. The only word they can form is QUIZ.

(b) **Between 111144 and 31114.** The words are in alphabetical order, except that O is at the end, rather than between N and P.

(c) The words are: A, BE, CUE, DUMB, ERROR, FINISH, GRADUAL, HEAVIEST, IGNORANCE, JUXTAPOSED, KNOWLEDGE, LEVITATE, MILITIA, NOTIFY, PEONY, Q--Z, RAW, SO, O. The hints are the letter frequencies of some of the words: 2-12-9-2-9-12-4-6/HEAVIEST & 4-12-2-9-6-9-6-12/LEVITATE, 3-6-9-4-4-9-4/GRADUAL & 2-9-4-9-6-9-9/MILITIA, 12-6-6-8-6/ERROR & 1-12-8-6-2/PEONY.

100.

(a) The Dark Side of the Moon – Pink Floyd

(b) Bat Out of Hell – Meat Loaf

(c) Misplaced Childhood – Marillion

(d) Nevermind – Nirvana

(e) Tubular Bells – Mike Oldfield

(f) AM – Arctic Monkeys

(g) Abbey Road – The Beatles

(h) Xscape – Michael Jackson

(i) The Circus – Take That

(j) London Calling – The Clash

(k) Aladdin Sane – David Bowie

(l) The Velvet Underground & Nico

(m) Rumours – Fleetwood Mac

(n) Artpop – Lady Gaga

(o) Automatic for the People – REM

(p) A Rush of Blood to the Head – Coldplay

(q) Led Zeppelin IV – Led Zeppelin

(r) Appetite for Destruction – Guns N' Roses

101.

14. Title represents Poetry. The missing words are Haiku, Limerick and Sonnet which have 3, 5 and 14 lines respectively.

102.

GROWL (TIGER) fits between NINETY (TEN) and SCHOOLBOY (TOM)

JENNY	MAC	SKIM	LITOVSK	HEAD
anydots	*avity*	*bleshanks*	*Brest*	*Bridge*
TIN	GARDENER	ME	DESPERATE	DENT
Can	*Capability*	*Centi*	*Dan*	*Deca*
SION	MUSTARD	SPIRIT	RUMP	DOC
Deci	*Dijon*	*Dunkirk*	*elteazer*	*Emmett*
MINE	CEILING	DETECTIVE	PM	O'FLAHERTY
Exa	*Fan*	*Father*	*Gordon*	*Half*
GILLIE	WIRELESS	ESSE	IRON	PHONE
John	*Lan*	*Lyon*	*Man*	*Mega*
DOT	CENT	UNSINKABLE	VICTORIA	PESHWARI
Micro	*Milli*	*Molly*	*Mrs*	*Nan*
REAGAN	BOT	SOLO	WORK	LENINIST
Nancy	*Nano*	*Napoleon*	*Nice*	*Nine*
MIST	MUNG	ETHENOL	BUST	SAUCE
offelees	*ojerry*	*One*	*opher*	*Pan*
HILTON	STRONG	LILLIPUTIAN	FACE	NOSE
Paris	*Patience*	*Pi*	*Poker*	*Rummy*
GENEVESE	MARXIST	DRAGON	MONKEY	ARC
Seven	*Six*	*Snap*	*Spider*	*Tan*
NINETY	**GROWL**	SCHOOLBOY	LAUTREC	GROWTH	LING
Ten	***Tiger***	*Tom*	*Toulouse*	*Two*	*Whist*

Cats in Old Possum's Book of Practical Cats by T.S. Eliot:
JENNY*anydots*, MAC*avity*, SKIM*bleshanks*, RUMP*elteazer*, (Mr.)
MIST*offelees*, MUNG*ojerrie*, BUST*opher* (Jones), **GROWL*tiger***

French towns/cities: *Brest*-LITOVSK, *Dijon* MUSTARD, *Dunkirk*
SPIRIT, *Lyon*ESSE, *Nancy* REAGAN, *Nice* WORK, *Paris* HILTON,
Toulouse LAUTREC

Card games: *Bridge* HEAD, *Napoleon* SOLO, *Patience* STRONG,
Poker FACE, *Rummy* NOSE, *Snap* DRAGON, *Spider* MONKEY,
Whist LING

Three-letter words ending AN: TIN *can*, DESPERATE *Dan*,
CEILING *Fan*, WIRELESS *Lan*, IRON *Man*, PESHWARI *Nan*,
SAUCE *Pan*, ARC *tan*

Browns: GARDENER (*Capability*), DOC (*Emmett*), DETECTIVE
(*Father*), PM (*Gordon*), GILLIE (*John*), UNSINKABLE (*Molly*),
VICTORIA (*Mrs*), SCHOOLBOY (*Tom*)

SI prefixes: *centi*ME, *deca*DENT, *deci*SION, *exa*MINE, *mega*PHONE,
*micro*DOT, *milli*CENT, *nano*BOT

Contain numbers reversed: O'FLAHERTY (*half*), LENINIST
(*nine*), ETHENOL (*one*), LILLIPUTIAN (*pi*), GENEVESE (*seven*),
MARXIST (*six*), NINETY (*ten*), GROWTH (*two*)

103.

These are all anagrams.

Scorpion & Bee Wings = Bowie & Prince Songs

(a) Drive-In Saturday

(b) When Doves Cry

(c) Hallo Spaceboy

(d) Paisley Park

(e) Space Oddity

(f) Girls and Boys

(g) Never Let Me Down

(h) U Got the Look

(i) Ashes to Ashes

(j) Purple Rain

(k) Loving the Alien

(l) Sign 'O' the Times

104.

>SWEDEN. Each word is an anagram of the part of a name for the day of the week preceding 'day'.

105.

>**50350.** The fractions $^1/_{11}$, $^1/_{10}$, ..., have decimal expansions 0.09(999...), 0.10(000...), 0.1(111...), 0.1250(000...), 0.142857(142857...), 0.16(666...), 0.20(000...), 0.250(000), 0.3(333...), 0.50(000...) and the numbers after the decimal points before recursion starts are:
>
>>09 10 1 1250 142857 16 20 250 3 50

106.

>**3rd March**
>
>The key is:
>
>ALEXA NDRIA BABYL ONEPH ESUSG IZAHA LICAR NASSU SOLYM PIARH ODES
>
>i.e. the locations of the 7 Wonders of the Ancient World, ending with Rhodes – the home of the Colossus.
>
>The messages read:
>
>**Message 1:**
>FROM HQ TO SECTIONS TEN AND TWELVE COMMENCE ATTACK ON MARCH THIRD
>
>**Message 2:**
>FROM SECTION TWELVE TO HQ MESSAGE RECEIVED AND UNDERSTOOD

107.

>(a) **WRONGED** has its letters in reverse alphabetical order. The other words have their letters in alphabetical order.
>
>(b) **CHIPS** has its letters in alphabetical order. The other words have their letters in reverse alphabetical order.

(c) **PUMPKIN** can be typed with just the right hand on a keyboard, the others can be typed with just the left hand.

(d) **SWEATER** can be typed with just the left hand on a keyboard, the others can be typed with just the right hand.

(e) **GADFLY**. The first Roman numeral in it is 10 times the second. In the others it is a tenth.

108.

OBEY. All the themes are related to science fiction or fantasy.

10: X Men: ANGEL, BEAST, BISHOP, FORGE, GAMBIT, JUBILEE, MARROW, MIMIC, STORM, VULCAN

9: Deep Space Nine: Words following DEEP (starting with D, E, E, P) and SPACE (starting with S, P, A, C, E). DEEP: DIVE, ECOLOGY, END, PURPLE. SPACE: SHUTTLE, PORT, AGE, CADET, EXPLORATION

8: EXTERMIN8: Eight-letter words which start and end cyclically with the letters EXTERMIN: EPICALYX, XEHANORT, TORTOISE, ENCIPHER, REHOBOAM, MACARONI, IMPRISON, NUISANCE

7: Blake's 7: Poems by William Blake: INTRODUCTION, JERUSALEM, LONDON, MILTON, SPRING, TIRIEL, VALA

6: The Sixth Sense: M, NIGHT, SHY, AM, ALAN, FILM

5: Characters from The Fifth Element: CORNELIUS, DALLAS, DAVID, FOG, ZORG

4: The Fantastic Four: BRILLIANT, MARVELLOUS, SUPERB, WONDERFUL

3, 2, and 1: Characters from Star Wars: SEA 3 PEA OWE, ARE 2 DEE 2, OBEY 1 (Kenobi)

109.

INDEX. Each word is the nth word in a sequence. US Presidents (Madison 4th), planets in the solar system (Saturn 6th), the Tonic Sol-fa scale (Do 1st), periodic table (Nitrogen 7th), books of the Bible (Exodus 2nd). By taking the nth letter in the word you can spell out **INDEX**.

110.

Flag transformations, changing colours:

Bl = Black, B = Blue, G = Green, R = Red, W = White, Y = Yellow

(a)   City & District of St Albans, or Flag of Mercia

(b)   Orkney Islands

(c)   Palau

(d)   West Florida/Somalia

(e)   Gloucestershire

(f)   Italy + 1824 → The Alamo

(g)   Madagascar → Benin, or Hungary → Lithuania

(h)   Guinea ← → Mali

111.

21. The substitutions are:

Original	0 1 2 3 4 5 6 7 8 9 / × + −
Replacement	1 4 8 5 7 0 2 3 9 6 − / × +

13 × 7 = 91	became	45 / 3 = 64
99 + 5 = 104	became	66 × 0 = 417
26 − 8 = 18	became	82 + 9 = 49
18 / 2 = 9	became	49 − 8 = 6

Hence

15 × 4 = 60	became	40 / 7 = 21

112.

The pairs form the names of countries when pronounced together. Pairs are:

CROW-ASIA (Croatia), SIR-BEER (Serbia), BELL-EASE (Belize), BREW-NIGH (Brunei), MALAISE-EAR (Malaysia), MONARCH-HOE (Monaco), JAW-JEER (Georgia), MOULD-OVER (Moldova), QUEUE-WEIGHT (Kuwait), VENICE-WHALER (Venezuela)

113.

Since they have a negative goal difference, we know that City lost at least one game, and have at most 6 points. With a goal difference of 0 we know that Wanderers have at least 3 points.

The only way a team can finish top (or bottom) with exactly 4 points is for all four teams to have exactly 4 points – this cannot have happened here, since City's goal difference is worse than Wanderers'. Hence Wanderers have exactly 3 points, and City have at least 5. But as City have lost a game, they cannot have exactly 5 points, so must have exactly 6 (WWL), with Wanderers having LLW. Thus Rovers and United between them have 2 wins and 2 losses against the other two teams.

As United only scored 1 goal, their goal difference is at most 1, so the total for City, United and Wanderers is at most 0; thus Rovers cannot have a negative goal difference, so (as they are below City) they must have less than 6 points, as must United.

Thus neither of them can have 2 wins, so they must each have beaten one of City and Wanderers and lost to the other, and the game between them must have been a draw.

With two wins and a loss but a negative goal difference, City's loss must have been by at least 3 goals, so must have been against Wanderers. Because they lost a game each, we know that United and Rovers each conceded at least 1 goal, and goal difference constraints then show they must each have conceded exactly 1 goal. Starting with United, this allows us to deduce all the scores.

Team	Points	Goals For	Goals Against	Goal Difference
City	6	5	6	−1
Rovers	4	2	1	1
United	4	1	1	0
Wanderers	3	6	6	0

Rovers	2	*v*	0	Wanderers
City	1	*v*	0	United
United	1	*v*	0	Wanderers
City	1	*v*	0	Rovers
United	0	*v*	0	Rovers
Wanderers	6	*v*	3	City

114.

(a) The words are examples of animals, vegetables and minerals – starting with letters A–H – enciphered with keys ANIMAL, VEGETABLE and MINERAL:

Sets are:

ANIMAL: ANT/AKT, BAT/NAT, COATI/IOATE, DOLPHIN/MOHPDEK, ELEPHANT/LHLPDAKT, FERRET/BLRRLT, GOOSE/COOSL, HYENA/DYLKA

VEGETABLE: ASPARAGUS/VQNVPVLSQ, BURDOCK/ESPTMGH, CABBAGE/GVEEVLA, DAIKON/TVDHMK, ENDIVE/AKTDUA, FENNEL/BAKKAI, GARLIC/LVPIDG, HORSERADISH/CMPQAPVTDQC

MINERAL: ANHYDRITE/MJBYEQCTR, BERYL/IRQYG, CYMRITE/NYHQCTR, DIAMOND/ECMHKJE, EMERALD/RERJCTR, FLUORITE/AGUKQCTR, GEORGEROBINSONITE/LRKQLRQKICJSKJCTR, HALITE/BMGCTR

Cipher alphabets are:

Plain: ABCDEFGHIJKLMNOPQRSTUVWXYZ
Cipher1: ANIMLBCDEFGHJKOPQRSTUVWXYZ
Cipher2: VEGTABLCDFHIJKMNOPQRSUWXYZ
Cipher3: MINERALBCDFGHJKOPQSTUVWXYZ

(b) The digraphs are the first and last letters of imperial units of length, volume and weight:

Length: INCH/IH, LINK/LK, FOOT/FT, YARD/YD, ROD/RD, CHAIN/CN, FURLONG/FG, MILE/ME,

Volume: GILL/GL, PINT/PT, QUART/QT, POTTLE/PE, GALLON/GN, PECK/PK, KENNING/KG, BUSHEL/BL

Weight: GRAIN/GN, DRAM/DM, OUNCE/OE, POUND/PD, STONE/SE, QUARTER/QR, HUNDREDWEIGHT/HT, TON/TN

115.

The names are anagrams of the members of Little Mix:

Jenny Soles – Jesy Nelson; Jilted War Hall – Jade Thirlwall; Congenial Pink Hen – Leigh-Anne Pinnock; Weirder Spread – Perrie Edwards

116.

The answers are all sets of 4 which have been summed (using A=1, B=2, etc, modulo 26) down the columns:

	(a)	(b)	(c)
	NORTHERNIRELAND	PESTILENCE	MATTHEW
+	SCOTLAND	FAMINE	MARK
+	ENGLAND	DEATH	LUKE
+	WALES	WAR	JOHN
=	FDWBKRHQIRELAND	TIVUCPENCE	SIBUHEW

117.

R	S	M	S	H	A	T	C	I
I	T	H	M	C	R	S	S	A
A	C	S	T	S	I	R	M	H
C	M	A	R	T	S	H	I	S
H	S	R	I	A	C	S	T	M
S	I	T	H	M	S	C	A	R
S	R	I	C	S	M	A	H	T
M	H	S	A	R	T	I	S	C
T	A	C	S	I	H	M	R	S

118.

BE.

F<u>IRST</u> = <u>G</u>OL<u>D</u>, <u>S</u>ECON<u>D</u> = <u>S</u>ILVE<u>R</u>, <u>T</u>HIR<u>D</u> = <u>B</u>RONZ<u>E</u>

119.

(a) Kth number is next value after K with same number of letters in name as K. So the sequence continues ..., 16, 70, **21, 19, 41,** ...

(b) C, CC, CCC, etc. Roman numerals in alphabetical order
354 = CCCLIV. Then CCCLIX = **359**, CCCLV = **355**, ...,
938 = CMXXXVIII. Then CV = **105**

(c) Birth states of US presidents, back from Trump
Richard Nixon = **California**, Lyndon B Johnson = **Texas**

(d) 15145, 202315, 2081855, 6152118, ?, ?, ...
ONE, TWO, etc. using A=1, B=2, etc. So ONE = 15145
then FIVE = **69225**, SIX = **19924**

(e) Map the QWERTYUIOP row on a keyboard to the digits in the row above in the standard way, using Q=1, W=2, ..., O=9, P=0. If a character doesn't appear in that row, use '.' instead. The sequence is the result of mapping ZERO, ONE, TWO, etc:

```
ZERO, ONE, TWO, THREE, FOUR, FIVE, SIX
.349, 9.3, 529, 5.433, .974, .8.3, .8.
```

120.

(a) 26830 Possible Games of Tic Tac Toe

5 Toes on a Foot

3 Feet in a Yard

1 Yard Of Ale is about 2.5 Pints

(b)
330	Frequency of E
370	Frequency of F Sharp
415	Frequency of G Sharp
440	Frequency of A
494	Frequency of B
554	Frequency of C
622	Frequency of D Sharp

Plain   A B C D E F G H I J K L M N O P Q R S T U V W X Y Z
Cipher E M A J O R S C L B D F G H I K N P Q T U V W X Y Z

(c) This represents the lines of the poem 'One, Two, Buckle My Shoe'

Plain   0 1 2 3 4 5 6 7 8 9
Cipher 1 5 4 2 0 3 8 6 9 7

(d) The A has 51 E, 21 H and 65 U

Plain   A B C D E F G H I J K L M N O P Q R S T U V W X Y Z 0 1 2 3 4 5 6 7 8 9
Cipher P O L Y H E D R A B C F G I J K M N Q S T U V W X Z 1 6 5 2 8   0   7

The tetrahedron has 4 faces, 6 edges and 4 vertices
The cube has 6 faces, 12 edges and 8 vertices
The octahedron has 8 faces, 12 edges and 6 vertices
The dodecahedron has 12 faces, 30 edges and 20 vertices
The icosahedron has 20 faces, 30 edges and 12 vertices

121.

K, S, T, V. The sequence is that of letters that start the names of multiple states of the USA. M and N start 8 names each; A, I and W start 4 names each; C and O start 3 names each; and K, S, T and V start 2 names each.

122.

TOMSK has replaced BUNGO in the names of other Wombles:

CHoLET, oRInoCo, buLgARIA; TobERMoRY, WELLIngTon

123.

All-America. The last words of each paragraph are the nicknames of US States, whose abbreviations form the word:

Cotton	Alabama	AL
Pelican	Louisiana	LA
Pine tree	Maine	ME
Ocean	Rhode Island	RI
Golden	California	CA

124.

USAGE, SUSTAIN, JOUST, TORUS, CAMPUS (inserting US at positions 0, 1, 2, 3, 4)

125.

EXAMPLE gives us 4-letter NATO phonetic alphabet words down the columns.

```
EXAMPLE
CRLIAIC
HAFKPMH
OYAEAAO
```

126.

    (a) **W**, as revealed by tracing out ABCDEFGHI.

        ANTIPAST**I**
        A**B**ERNETHY
        AR**C**LEN**G**TH
        ACI**D**I**F**IED

    (b) **δ**, as revealed by ALFA, BRAVO, CHARLIE

        LE**AFL**ADEN
        ABSEILING
        TRILLIONS
        UN**AVO**IDED
        RE**FRA**CTED
        MILKSHAKE
        GI**LRA**VAGE

    (c) **X**, as revealed by tracing out the X's in (Charli) XCX, (Battle) AXE, and (the last King Louis of France) (Louis) XIX:

        XC**X**
        A**X**E
        XI**X**

    (d) $\sqrt{2}$, as revealed by tracing out 1.41421356 and all the 2's:

        9.07184**356**
        3.5671**1**2**2**8
        4.433**22**112
        1.65**4**54321
        5**.**3**1**974286
        4.**4**443**2222**

127.

The words contain the numbers 1 to 5 in various languages.

(a) Teleph**one**, Trus**two**rthy, **Three**some, Bal**four**, **Five**pence (English)

(b) **Veins**, Rosen**zwei**g, An**drei**, Oli**vier**, **Funf**air (German)

(c) **Michi**gan, Pho**eni**x, **San**ta Ana, Wa**shi**ngton, Ore**gon** (Japanese)

128.

Bands and their albums:

Aerosmith	Get a Grip, Rocks, Toys in the Attic
Boyzone	A Different Beat, Said and Done, Where We Belong
Coldplay	A Head Full of Dreams, A Rush of Blood to the Head, Mylo Xyloto
Devo	Q: Are We Not Men? A: We Are Devo!, Shout, Total Devo
Eagles	Hotel California, One of These Nights, Long Road Out of Eden
Foreigner	Agent Provocateur, Can't Slow Down, Inside Information
Genesis	A Trick of the Tail, Selling England by the Pound, The Lamb Lies Down on Broadway
Heart	Bad Animals, Dog & Butterfly, Dreamboat Annie
INXS	Kick, Listen Like Thieves, Shabooh Shoobah
Journey	Evolution (or Escape or Eclipse), Frontiers, Infinity
Kansas	Point of Know Return, Song for America, The Prelude Implicit
Lush	Scar, Split, Spooky

Motörhead	Ace of Spades, Bad Magic, Clean Your Clock (live album)
Nirvana	Before We Ever Minded (compilation album), In Utero, Nevermind
Oasis	Be Here Now, Don't Believe the Truth, Dig Out Your Soul
Poison	Flesh & Blood, Look What the Cat Dragged In, Native Tongue
Queen	A Night at the Opera, Hot Space, Made in Heaven
Ramones	Acid Eaters, Rocket to Russia (or Road to Ruin), Too Tough to Die
Sweet	Cut Above the Rest, Give Us a Wink, Off the Record
Technotronic	Body to Body, Pump Up the Jam, Recall
Ultravox	Quartet, Systems of Romance, Vienna
Visage	Beat Boy, Demons to Diamonds, Hearts and Knives
Wham!	Fantastic, Music from the Edge of Heaven, Make It Big
XTC	Drums and Wires, Oranges & Lemons, Skylarking
Yes	Close to the Edge, Going for the One, Keys to Ascension
Zodiac	A Bit of Devil, Grain of Soul, Sonic Child

129.

(a) **NATO phonetic alphabet**. T3215 denotes the 3rd, 2nd, 1st and 5th letters (NATO) of T/TANGO, A34 the 3rd and 4th letters (PH) of A/ALPHA, and so on, spelling out the phrase.

(b) **The periodic table**. N3 denotes the 3rd letter (T) of N/NITROGEN, P6 the 6th letter (H) of P/PHOSPHORUS and so on, spelling out the phrase.

(c) **International vehicle registration codes.** V473 denotes the 4th, 7th and 3rd letters (INT) of V/VATICAN CITY, F6243 the 6th, 2nd, 4th and 3rd letters (ERNA) of F/FRANCE and so on, spelling out the phrase.

(d) **Seen in the even columns.**

TS<u>A</u>VO
BE<u>B</u>EK
BE<u>R</u>NE
ONE<u>C</u>O
DI<u>J</u>ON
IN<u>O</u>LA
AT<u>H</u>UR
THA<u>M</u>E
CE<u>R</u>NA
NE<u>U</u>SS

(e) **Sieve of Eratosthenes,** an ancient algorithm for finding prime numbers. Regarding the sequence as a string of letters, this phrase is spelled out by striking out the non-prime-numbered letters.

(f) **Catholic popes.** The 4th Pope's name is CLEMENT I and the first (Ist) letter of **C**LEMENT is C; this continues 6/ **A**LEXANDER I, 73/**T**HEODORE I, 61/JO**H**N III, 55/ **B**ONIFACE II, 22/**L**UCIUS I, 24/S**I**XTUS II, 16/**C**ALLISTUS I, 10/**P**IUS I, 56/JO**H**N II, 60/**P**ELAGIUS I, 63/P**E**LAGIUS II, 7/**S**IXTUS I

(g) **Require top typewriter row.** Replacing each number n by the nth letter in the string QWERTYUIOP spells out the phrase.

(h) **The last part of this question.** This phrase is spelled out by the 31st, 63rd, 59th, etc, letters in the text of this question.

130.

Emmanuel. These are the surnames of famous guitarists, and their first letters are the sequence of strings on a guitar:

Tommy Emmanuel, Chet Atkins, Bob Dylan, David Gilmour, Jeff Beck – Tommy Emmanuel (or Duane Eddy)

131.

TD, CE, F. This was Quality Street Snooker! The letters represent the names of chocolates inside wrappers of the appropriate colour. Only Red, Yellow, Green, Brown, Blue and Pink exist in the game, hence the maximum break is only 125. The colours are:

(1) Red   Strawberry Delight

(2) Yellow   Caramel Swirl

(3) Green   Green Triangle

(4) Brown   Toffee Deluxe

(5) Blue   Coconut Eclair

(6) Pink   Fudge

132.

Words alternate between categories and examples.

(a) Planet of the solar system, **Venus**, Bananarama No. 1 single

(b) Humphrey Bogart movie, **Casablanca**, Moroccan city

(c) Brown, **Ivy League university**, **Columbia**, Space shuttle

(d) Madonna No. 1 single, **Frozen**, **Disney movie**, 101 Dalmatians

(e) Genesis album, **Foxtrot**, **Ballroom dance (or NATO phonetics)**, **Tango**, Fizzy drink

(f) Sliver, **Nirvana song**, **Lithium**, **Element**, Silver

(g) Joshua Reynolds's painting, **The Age of Innocence**, **Daniel Day Lewis film**, **Lincoln**, US state capital

Other paths are possible, e.g. (g) can be linked through characters from A Midsummer Night's Dream.

133.

(a) Pairs form powers of 2:

1–28, 2–56, 5–12, 10–24, 204–8, 40–96, 81–92, 163–84, 32–768, 655–36

(b) Pairs multiply together to form numbers with repeated digits:

$1 \times 9 = 9$
$4 \times 22 = 88$
$21 \times 37 = 777$
$33 \times 202 = 6,666$
$205 \times 271 = 55,555$
$462 \times 962 = 444,444$
$717 \times 4649 = 3,333,333$
$1606 \times 13837 = 22,222,222$
$333 \times 333667 = 111,111,111$

(c) Pairs form 42 (the answer) in different ways:

$3 \times 14 = 42$
$294 / 7 = 42$
$10 + 32 = 42$
$142 - 100 = 42$

134.

Across		Down	
3	Dressing	1	Diggy
5	Transversely	2	Firefox
7	Melancholy	3	Despondent
8	Opposite	4	Drink
11	Over	6	Fed up
12	Morose	9	Out of order
13	Jaded	10	Beyond

135.

67. The functions k, w, i and z are defined by:

k(n) = next prime after n

w(n) = number formed from reversing the digits of n

i(n) = sum of digits in n

z(n) = square of n

So kwiz(23) = k(w(i(z(23)))) = k(w(i(529))) = k(w(16)) = k(61) = 67

136.

JW. Jodie Whittaker is the latest in the sequence of actors to play The Doctor in Doctor Who.

137.

GS. Gareth Southgate is the next England football manager in the sequence.

138.

Chestnuts roasting on an open fire
Come, they told me, pa rum pum pum pum
Dashing through the snow
From Heaven above to Earth I come
Good King Wenceslas looked out
Hark how the bells
Have yourself a merry little Christmas
Here we come a-wassailing
I'm dreaming of a white Christmas
It's Christmas time, there's no need to be afraid
Just hear those sleigh bells jingling
O holy night, the stars are brightly shining
On the first day of Christmas
Sleigh bells ring, are you listening?
So this is Christmas
The child is a king, the carollers sing
They said there'll be snow at Christmas
Welcome to my Christmas song

139.

The answer to each clue is a letter from the NATO phonetic alphabet.

(a)   The nymph was in love with NARCISSUS. (Echo)

(b)   The poem was written by THOMAS HOOD. (November)

(c)   The country is one of the BRIC economies. (India)

(d)   The car is manufactured by VOLKSWAGEN. (Golf)

(e)   The schoolboy novel is by P. G . WODEHOUSE. (Mike)

(f)   The particle is a HELIUM NUCLEUS. (Alpha)

140.

Gloucester. Translating each into Welsh and extracting the nth letter, we obtain Caerloyw, which is Welsh for Gloucester.

CAERDYDD
CA**S**TELLNEDD
ABE**R**TAWE
CAE**R**GYBI
YGEL**L**I
PENYB**O**NT
CASNEW**Y**DD
TREFALD**W**YN

141.

These are the coordinates of Cheltenham's twin towns: Annecy (France), Cheltenham (USA), Göttingen (Germany), Sochi (Russia), Weihai (China).

142.

Each is an anagram of a musical tempo.

GOTH ALERT – LARGHETTO
AN INN TOAD – ANDANTINO
MORE TOAD – MODERATO

MAIN OAT – ANIMATO
POSTER – PRESTO
ITS PROMISES – PRESTISSIMO

143.

(a) All the numbers are prime, but the numbers on the left also have a prime number of letters when written out in English. The number on the right has a square number of letters. TWO, 3; THREE, 5; TWENTY THREE, 11; ONE HUNDRED AND NINE, 17; ONE HUNDRED AND THIRTY NINE, 23; ONE THOUSAND ONE HUNDRED AND THREE, 29; ONE THOUSAND THREE HUNDRED AND THREE, 31; ONE THOUSAND THREE HUNDRED AND TWENTY SEVEN, 37; ELEVEN THOUSAND SEVEN HUNDRED AND SEVENTY SEVEN, 41; THIRTEEN THOUSAND EIGHT HUNDRED AND SEVENTY THREE, 43 : ONE THOUSAND TWO HUNDRED AND SEVENTY SEVEN, 36

(b) To the left of the colon, each of the numbers in a pair (a, b) is prime, as are the concatenations ab and ba. [1123, 2311], [1318, 1913], [1783, 8317], [1979, 7919], [1997, 9719], [2347, 4723], [2971, 7129], [3767, 6737], [3779, 7937], [4397, 9743]. To the right, each of (a, b) is prime, but neither of ab and ba are prime: [4771, 7147].

(c) Interpreting the quadruples as atomic numbers, those to the left of the colon spell out words when each atomic number is replaced by that element's symbol: H-O-NE-Y, HE-C-TA-RE, LI-TI-GA-TE, BE-B-O-P, B-Y-LA-W, C-AS-I-NO, N-O-SE-Y, O-S-I-ER, F-I-ND-ER, NE-S-TL-ES. The quadruple to the right spells out a reversed word: NA-I-CE-RG / GR-EC-I-AN.

144.

66. Each three-letter word is an anagram of the first three letters of the country to win the FIFA World Cup in the corresponding year. England won in 1966.

145.

> EGYPT, SCANDINAVIA. In the game of Risk there are 42 territories in 6 continents. The full list consists of the territories with the letters in its continent removed (e.g. GREAT BRITAIN is in EUROPE and, having removed the common letters ER, becomes GAT BITAIN; ONTARIO is in NORTH AMERICA and, having all its letters in common with its continent, is represented by '–'). The missing items are EGYPT and SCANDINAVIA, the 2 territories that have no letters in common with the continents (AFRICA and EUROPE).

146.

> URBAN is an anagram of BRAUN, which is German for Brown. The other words are anagrams of French colours: GARNET – ARGENT (Silver); IRON – NOIR (Black); RIGS – GRIS (Grey); ROGUE – ROUGE (Red); SORE – ROSE (Pink)

147.

TV sitcoms and their prequels/sequels:

(a)   Are You Being Served → Grace & Favour

(b)   The Likely Lads → Whatever Happened to the Likely Lads?

(c)   Cheers → Frasier

(d)   Friends → Joey

(e)   Rock and Chips ← Only Fools and Horses → The Green Green Grass

(f)   Fresh Fields → French Fields

(g)   Seven of One (Part 2): Prisoner and Escort → Porridge → Going Straight

148.

ICROSOF made the BO. By adding a letter to the beginning and end of each capitalized word, the statement reads 'If NINTENDO made the WII and SONY made the VITA, who made the XBOX?'

149.

Lengths of words in first lines of songs, using A=1, B=2, ...

(a) **Deck the Halls** – Deck the halls with boughs of holly, fa la la la la la la la la

(b) **We Wish You a Merry Christmas** – We wish you a merry Christmas, we wish you a merry Christmas, we wish you a merry Christmas, and a happy New Year

(c) **When a Child Is Born** – A ray of hope flickers in the sky, a tiny star lights up way up high, all across the land, dawns a brand new morn

(d) **Santa Baby** – Santa baby, slip a sable under the tree for me, been an awful good girl, Santa baby, and hurry down the chimney tonight

(e) **It's the Most Wonderful Time of the Year** – It's the most wonderful time of the year, with the kids jingle belling and everyone telling you be of good cheer

(f) **When Santa Got Stuck Up the Chimney** – When Santa got stuck up the chimney, he began to shout, you girls and boys won't get any toys if you don't pull me out

(g) **Christmas Alphabet** – C is for the candy trimmed around the Christmas tree, H is for the happiness with all the family

(h) **Santa Claus is Coming to Town** – You better watch out, you better not cry, better not pout, I'm telling you why, Santa Claus is coming to town

(i) **Rockin' Around the Christmas Tree** – Rockin' around the Christmas tree at the Christmas party hop, mistletoe hung where you can see every couple tries to stop

(j) **Once in Royal David's City** – Once in royal David's city stood a lowly cattle shed, where a mother laid her baby in a manger for his bed

150.

Dogberry (from Much Ado About Nothing). The question contains classes of the Order of the British Empire in ascending order of precedence: If you remeMBEr enjoying ROBErt Carlyle as Hamish MaCBEth, you'll love seeing Jim BroaDBEnt rise from his sicKBEd to play – DoGBErry.

151.

(a) ONE/ONE, TWO/DBJ, THREE/SPO, FOUR/HDT, FIVE/WRY, SIX/LFD, SEVEN/ATI, EIGHT/PHN, NINE/EVS, TEN/TJX

Using A=1, B=2, ... Z=26, and adding letter by letter:

ONE+ONE = 2×ONE = DBJ, ONE+ONE+ONE = 3×ONE = SPO, etc

(b) Similarly to (a), each cardinal ONE, TWO, ..., TEN can be added to itself letter by letter. 2520 (the least common multiple of 1, 2, ..., 10) can be expressed:

BFPFV = 360 × SEVEN
DLNNN = 840 × THREE
FPX = 1260 × TWO
HLLX = 504 × FIVE
JLVD = 630 × FOUR
OAUXH = 315 × EIGHT
TXTV = 280 × NINE
VLR = 252 × TEN
VXP = 2520 × ONE
XJR = 420 × SIX

152.

(a) Italy/Switzerland

(b) Ukraine/Poland (N is pointing S)

(c) Sweden/Finland

(d) Spain/Portugal (N is pointing W)

(e) Argentina/Chile

(f) Iraq/Saudi Arabia (N is pointing SE)

(g) Iran/Afghanistan

(h) North Korea/South Korea (N is pointing S)

(i) Ethiopia/Somalia (N is pointing E)

(j) Algeria/Libya

153.

They are in snooker-ball value order: (red) HEAD, (yellow) BELLY, (green) FINGERS, (brown) NOSE, (blue) BEARD, (pink) EYE, (black) LEG

154.

(a) **The Four Shire Stone** (51° 51.25'N, 001° 39.94'W)

33.22km from Warwick Shire Hall

(b) **The Beatles**: the given pieces match the up/down/same Parsons Code values for Yesterday, Strawberry Fields, A Hard Day's Night and Hey Jude respectively, up to the specified position

(c) Films and the Shakespeare plays that inspired them:

Forbidden Planet (The Tempest)
Ran (King Lear)
She's the Man (Twelfth Night)
West Side Story (Romeo and Juliet)
10 Things I Hate About You (The Taming of the Shrew)

(d) **Edgar Degas**

155.

(a) Nursery rhymes and folk tales: $(3\times3\times3) - 3 = 24 =$ Four-and-twenty **blackbirds** baked in a pie

(b) Formula One driver numbers: Kamui Kobayashi = 10, Felipe Nasr = 12 <u>or</u> Felipe Massa = 19, Kimi Raikonnen = 7, Esteban Gutierrez = 21 <u>or</u> Esteban Ocon = 31, Jolyon Palmer = 30, Jules Bianchi = 17, Rio Haryanto = 88, Lance Stroll = 18. This gives four possible sums, with three different answers:

$((10 \times 12) + (7 \times 21) + 30 - 17) / (88 - 18) = 4 =$ **Max** Chilton

$((10 \times 19) + (7 \times 21) + 30 - 17) / (88 - 18) = 5 =$ **Sebastian** Vettel

$((10 \times 12) + (7 \times 31) + 30 - 17) / (88 - 18) = 5 =$ **Sebastian** Vettel (again)

$((10 \times 19) + (7 \times 31) + 30 - 17) / (88 - 18) = 6 =$ **Nico** Rosberg

(c) Beethoven symphonies:

Symphony numbers: $((5 - 3) \times 7 / (4 - 2)) + 1 = 8 =$ **F major**

Opus numbers: $((67 - 55) \times 92 / (60 - 36)) + 21 = 67 =$ **C minor**

(d) This uses a standard UK keyboard, 1=!, 2=", 3=£, 4=$, 5=%, 6=^, 7=&, 8=*, 9=(and 0=).

Depending on how () and ! are interpreted, the following valid expressions can be derived:

$37 - \sqrt{76 + (45 + 42) / 3 + 51} - 56$ $= 37 - \sqrt{100}$ $= 27$ $=$ "&

$37 - \sqrt{76 + (45 + 42) / 3 + 5!} - 56$ $= 37 - \sqrt{169}$ $= 24$ $=$ "$

$37 - \sqrt{76 + 945 + 420 / 3 + 51} - 56$ $= 37 - \sqrt{1156}$ $= 3$ $=$ £

$37 - \sqrt{76 + 945 + 420 / 3 + 5!} - 56$ $= 37 - \sqrt{1225}$ $= 2$ $=$ "

156.

SIGH. The wordbox contains homophones of Greek letters where they exist, other than for PSI.

In the wordbox are: BEATER, EATER, CAPPER, MEW, KNEW, PIE, ROW, TORE, FIE.

157.

Ellistown – starting with the first letter of Ellistown (E), the subsequent directions are those moved on a phone keypad to spell out the location (a '–' means no move)

Nunnykirk – as above, starting with N

Stivichall – as above, starting with S

Woodmancott – as above, starting with W

158.

VIOLIN, VIOLA, CELLO, DOUBLE BASS, the string section of an orchestra.

159.

(a) **10th.** YESTERDAY is the 1st word in its lyrics; PENNY LANE (preceded by IN) starts at the 2nd word; A HARD DAY'S NIGHT (preceded by IT'S BEEN) at the 3rd; NOWHERE MAN (preceded by HE'S A REAL) at the 4th and SUN KING (preceded by AH – HERE COMES THE) at the 5th. OCTOPUS'S GARDEN is preceded by I'D LIKE TO BE UNDER THE SEA IN AN.

(b) **8.** The first word (I) in the lyrics of NORWEGIAN WOOD has 1 letter; that of YELLOW SUBMARINE (In) 2; REVOLUTION (You) 3; BACK IN THE USSR (Flew) 4; DRIVE MY CAR (Asked) 5. The first word in the lyrics of THE BALLAD OF JOHN AND YOKO (Standing) has 8 letters.

160.

2, 2, 1, 1. These are the number of each of the body parts listed in the nursery rhyme Heads, Shoulders, Knees and Toes. The missing entries represent (2) eyes and (2) ears and (1) mouth and (1) nose.

161.

1	2	3	4	5	6	7	8	9	10	11	12	13	14	15	16	
				T			B								S	
		D		O		N	A				T				E	
	S	N	K	O		R	C	T	E	W	U			S	N	
W	H	I	C	H	D	U	K	E	M	A	R	C	H	E	D	?
I	O	W	A	S	N	T		S	O	S	N	O	S	T		
N	O		B		E				C	H		M	A			
D	T				S							E	W			

The clues lead to two forms of a verb, and this should be entered in the grid in the appropriate direction.

1, 3 Wind Down, Wind Up

2, 5 Shoot Down, Shoot Up

4, 8 Back Up, Back Down

6, 16 Send Up, Send Down

7, 12 Turn Up, Turn Down

9, 15 Set Up, Set Down

10, 13 Come Up, Come Down

11, 14 Wash Down, Wash Up

Across the middle reads: Which Duke Marched?

To which the answer is The Grand Old Duke of York.

162.

4040401040002070306090203010901090203030806020407

×

80000000000000000000000000000000000000000000000011

163.

These are all the lengths of names:

(a) **U=6.** The Muses: Clio, Erato, Thalia, Euterpe, Calliope, Melpomene, Polyhymnia, Terpsichore and **Urania**

(b) **P=4.** The Beatles. John, Ringo, George and **Paul**

(c) **S=5.** Narnia kings and queens: Lucy, Peter, Edmund and **Susan**

(d) **DC=NC=15.** The Simpsons actors: Hank Azaria, Julie Kavner, Harry Shearer, Yeardley Smith, **Dan Castellanata and Nancy Cartwright**

164.

LISBON, 3.

The distances are edit distances, i.e. the number of changes (removal, insertion or change of one letter) needed to get from one capital to the next:

CAIRO – PAIRO – PAIRS – PAIS – PARIS (4)
PARIS – MARIS – MADRIS – MADRID (3)
MADRID – BADRID – BEDRID – BERRID – BERLID – BERLIN (5)
BERLIN – DERLIN – DURLIN – DUBLIN (3)

LONDON – LINDON – LISDON – LISBON (3)

165.

Change the first and last letter of each word to create pairs. Pairs are:

BEGINS, PINNIPED – REGINA, WINNIPEG (Canadian capitals)

BRANCH, STALL – FRANCE, ITALY (European countries)

DELIUS, VENOM – HELIUM, XENON (Noble gases)

GAWAIN, PAINT – HAWAII, MAINE (US States)

GUMMO, TOKEN – RUMMY, POKER (Card games)

HOPED, TRUMPS – DOPEY, GRUMPY (Dwarfs)

MURINE, PUTTY – TURING, TUTTE (Bletchley Park codebreakers)

NATURE, PARTY – SATURN, EARTH (Planets)

NORTH, SHIRTS – FORTY, THIRTY (Multiples of 10)

OASIS, PORTRAY – BASIC, FORTRAN (Computer languages)

166.

Answer. The words have appropriate compass directions in their centres:

ruNWay liNEar

anSWer wiSEly

167.

The missing words are BASIC instructions and can be found on old Sinclair keyboards (such as the ZX Spectrum), on the same keys as the letters in Enigma.

(a)   The Athens band is REM.

(b)   The clothes retailer is NEXT.

(c)   The deductible VAT is INPUT.

(d)   The GOTO islands are part of Nagasaki Prefecture.

(e)   <U+1D110> denotes a PAUSE.

(f)   The Paul McCartney album is NEW.

168.

Film titles involving US states:

(a)   The Hotel **NEW HAMPSHIRE**

(b)   My Own Private **IDAHO**

(c)   The **TEXAS** Chain Saw Massacre

(d) **OKLAHOMA!**

(e) **MISSISSIPPI** Burning

(f) The **KENTUCKY** Fried Movie

(g) The Prince of **PENNSYLVANIA**

(h) Barr**IO WA**rs

169.

The words contain the numbers 1 to 5 in various languages.

(a) Frank**en**stein, **Tom**, **Tre**ble, **Fire**fly, **Fem**ale (Norwegian)

(b) Pin**yin**, **Er**ror, Nis**san**, Pep**si**, Beo**wul**f (Mandarin)

(c) **Satu**rn, **Dua**lity, **Litiga**te, **Empa**thy, **Clima**te (Malay)

(d) Wom**bat**, **Bi**t, **Hiru**din, Glau**coma**, **Bost**on (Basque)

170.

To get started with this puzzle we observe that since United have no points, every team beat them, so Rovers's one goal must have been from a 1-0 win against United, and the most points they can possibly have is 5. However, since every team beat United, neither City nor Wanderers have won another game either (or else they would be on at least 6 points and above Rovers). Since they can't have won, and they can't have lost, they must have drawn, and so they have 5 points. Finally, we use the tiebreaker rules to deduce that neither City nor Wanderers scored more goals than Rovers, and since they scored at least 1 to win against United, all the other games must have been 0-0 draws. We conclude that Rovers must be on the top of the table as a result of fair play or the drawing of lots.

Team	Points	Goals For	Goals Against	Goal Difference
Rovers	5	1	0	1
City	5	1	0	1
Wanderers	5	1	0	1
United	0	0	3	−3

City	1	v	0	United
Rovers	0	v	0	Wanderers
Wanderers	1	v	0	United
Rovers	0	v	0	City
Wanderers	0	v	0	City
United	0	v	1	Rovers

171.

The rainbow. Colours are missing from each set of letters:

baRED dOwnRANGE plaYfELLOW aGREEmeNt BLUndEr INDIGestiOn VIOLEnT

172.

The words can be paired and anagrammed to form the Roman names of British towns.

COBRA, EMU – Eboracum (York)

GEM, LUV – Glevum (Gloucester)

MICRO, UNI – Corinium (Cirencester)

MILD, UNION – Londinium (London)

MIME, UVULAR – Verulamium (St Albans)

173.

PENNILESS. The other words are formed by taking rivers and adding letters to the middle to form new words. LOgfIRE, OUthouSE, SErpentINE, TImBER, TReatmENT, TumbleWEED. PenNILEss adds letters around a river to form a new word.

174.

Rarity and Fluttershy. The four clues give the answers Twilight Sparkle, Rainbow Dash, Applejack and Pinkie Pie. Rarity and Fluttershy complete the Mane Six equines from My Little Pony.

175.

(a) **4.10.** Reverse each pair to read: Sack of Rome by the Visigoths (in AD 410).

(b) **3.54.** Write the Roman numerals underneath the groups and subtract using I=1, V=5, X=10.

```
  U  IF UJN FS  T  BJ XUI WFFG  JP  D  IG  YVS
- I  II III IV  V  VI VII VIII  IX  X  XI  XII
─────────────────────────────────────────────
  T  HE TIM EN  O  WI STH REEF  IF  T  YF  OUR  =  The time now
                                                    is three fifty-
                                                    four
```

(c) **12.20.** Each group represents one letter. Add the Roman numerals to the groups using A=1, B=2, etc

```
  K  NN VVV CP  Z  IV XKK ANNN  VG  P  VK  APP
+ I  II III IV  V  VI VII VIII  IX  X  XI  XII
─────────────────────────────────────────────
  T  WW EEE LL  V  EE TTT WWWW  EE  N  TT  YYY  =  Twelve twenty
```

(d) **9.56.** Index each word by the length of the Roman number:

I	Sneeze
II	1Ombok
III	tuRkey
IV	oTters
V	Ivanka
VI	iNmate
VII	miTten
VIII	nagOya
IX	wAxing
X	Tsetse
XI	vOwels
XII	nuZzle

This spells: Sort into A to Z. Do this with the words, and index them again by the length of the Roman number:

I	Inmate
II	iVanka

```
III    loMbok
IV     mItten
V      Nagoya
VI     nUzzle
VII    otTers
VIII   sneEze
IX     tSetse
X      Turkey
XI     vOwels
XII    waXing
```

This spells: IV minutes to X.

(e) **19.55.**

The question was encoded using the alphabet:

Plain: ABCDEFGHIJKLMNOPQRSTUVWXYZ
Cipher: IIIIIIIVVVIVIIVIIIIXXXIXII

The question reads:

One, Two, Three o'clock, Four o'clock rock, Five, Six, Seven o'clock, Eight o'clock rock, Bill Haley and His Comets, United Kingdom, Hit, When?

176.

At the time of publication you can reach 501 in 3 darts in three different ways: (M275, M181, M45), (M271, M181, M49) or (M271, M180, M50).

177.

100. The set is KANGAROO, SHAMPOO, TATTOO, IGLOO, ALOO, 100. We are celebrating 100 years – but 00 isn't quite the same as OO.

178.

Motorway service stations:

(a) Sarah **Winchester** (M3)

(b) **Keele** (M6) or **Chester** (M56)

(c) **Michael Wood** (M5)

(d) **Fleet** Street (M3)

(e) **Magor** (M4)

(f) Rachael **Stirling** (M9/M80)

(g) **Birch** (M62)

(h) **Heston** Blumenthal (M4)

179.

Each clue has 2 consecutive letters associated with it.

BA – Qualification; DC – Washington; FE – Iron; HG – Wells;
JI – -k (Quaternion); LK – Lunar lander; NM – Nevermind;
PO – Tellytubby; RQ – CO_2/O_2 (respiratory quotient); TS – Eliot;
VU – Déjà; XW – Original Starfighter; ZY – Sky Airlines

180.

auar, e

The letters are the overlaps between the signs of the Zodiac and the
months in which they start:

ARies, tAuRus, geMini, caNcEr, Leo, virGo, liBRa, sCORpiO,
sagittaRius, CapRicorn, AqUARius, piscEs

mARch, ApRil, May, juNE, juLy, auGust, septemBeR, OCtObeR,
novembeR, deCembeR, jAnUARy, fEbruary

181.

(a) **102.** The other numbers are the lowest numbers that, when spelled out, begin and end in N, T and E respectively: 19/NINETEE_N_, 28/_T_WENTY EIGH_T_, 81/_E_IGHTY ON_E_. The only other letter which can begin and end a number is O, and the lowest such number is 102/_O_NE HUNDRED AND TW_O_.

(b) **4000000.** The other numbers are the lowest numbers that, when spelled out, have the letters E, W, H, I, N respectively as both the second and penultimate letter: 7/S_E_V_E_N, 22/T_W_ENTY T_W_O, 38/T_H_IRTY EIG_H_T, 56/F_I_FTY S_I_X, 101/O_N_E HUNDRED AND O_N_E. The only other letter that can occupy both these positions is O, and the lowest such number is 4000000/F_O_UR MILLI_O_N.

(c) ONE MI_LL_ION _N_I_N_ETY EIGH_T_ _T_HOUSAND AND TH_RE_E is the smallest number with 4 different pairs of identical letters.

182.

PANNIER – it has a musical (ANNIE) in it running forwards, the others have EVITA, NINE, HAIR, RENT and CATS backwards.

183.

Russia. It's derived from Churchill's 1939 quote '. . . Russia. It is A RIDDLE (anagram of LED RAID) wrapped in A MYSTERY (anagram of MAY TYRES) inside AN ENIGMA (anagram of MEN AGAIN) . . .'

184.

0. These are the first numbers that both contain the nth letter of the alphabet and are a multiple of n, starting with 26 (the last letter of the alphabet.) The next term is the 18th letter of the alphabet, R. You were probably tempted to answer 36, but congratulations if you spotted that ZERO also contains an R.

185.

DISCOVER. Tracing out a path through the letters in each site name draws each letter of the answer in turn, if you look from the direction indicated. The hills can be considered to be made of glass.

186.

WRY-RYE-YEW-EWE

187.

(1) 273. twO, seveN, threE

(2) 372, 732. End with 2

(3) 27, 1594323, 5559060566555523. Powers of 3, power ending in 3. ($27=3^3$, $1594323=3^{13}$, $5559060566555523=3^{33}$)

(4) 6, 15, 21, 18. Spell FOUR (A=1, B=2, etc)

(5) 5-digit numbers forming a 5×5 square

1	1	2	3	5
1	2	4	8	1
2	4	6	8	1
3	8	8	0	9
5	1	1	9	7

(6) 519, 1134, 1652158, 743665127, 34159739501, 37907741687. Sixth in a series, written in base 36($=6^2$): 519=EF, 1134=VI, 1652158=ZETA, 743665127=CARBON, 34159739501=FOXTROT, 37907741687=HEXAGON

(7) 13, 14, 17, 49, 823543, 1000006, 8000000000085. Seventh in a series: 13 (Fibonacci/odd), 14 (Composite/even), 17 (prime), 49 (square), 823543(7^7), 1000006 (7-digit number), 8000000000085 (alphabetically when spelled UK style Eight Billion and Eighty-Five)

(8) 24, 8, 160, 40, 56, 120, 144, 200. Divisible by 8, and when divided spell out CATEGORY

(9) 925, 946, 9627, 9729, 948437, 9447539, 94478533, 96335533, 995674663. Begin with 9, and correspond to a texted word: 925=YAK, 946=ZHO, 9627=XMAS, 9729=XRAY, 948437=ZITHER, 9447539=WHISKEY, 94478533=WHISTLED, 96335533=YODELLED, 995674663=XYLOPHONE

(10) 19, 37, 433, 721, 6121, 8011, 123211, 205030, 1330210, 4002103. Digits add up to 10

188.

The Man with the Golden Gun. They are James Bond films with successively 1, 2, 3, 4, 5 and 6 words in the titles. Additionally there is one film by each of the main six actors to have played James Bond.

189.

(a)   Oscar winners for Best Picture in prime-number years.

1931 Cimarron

1933 [no awards]

1949 Hamlet

1951 All About Eve

1973 The Godfather

1979 The Deer Hunter

1987 Platoon

1993 Unforgiven

1997 The English Patient

1999 **Shakespeare In Love**

2003 **Chicago**

2011 **The King's Speech**

2017 **Moonlight**

(b)   Numbers with exactly one of each vowel, e.g. TWO HUNDRED AND SIX

then **26000, 80000, 90000**

(c)   Sea, Hay, Char, Rye, Yes, Tea

Sounds like C-H-R-I-S-T, then -M-A-S = **Emmaus**

190.

Elizabeth. These are the second names of the Queen's great grandchildren from oldest to youngest:

Savannah Anne Kathleen Phillips

Isla Elizabeth Phillips

(Prince) George Alexander Louis

Mia Grace Tindall

(Princess) Charlotte Elizabeth Diana

(Prince) Louis Arthur Charles

Lena Elizabeth Tindall

191.

Jamaica. Some people also claim Sri Lanka, but the background to the lion is a shade of red.

192.

(a)   **ROOSTER.** WRY and MAURIE rhyme with Fry and Laurie, who portrayed Jeeves and Wooster, which rhyme with Peeves and Rooster.

(b)   **TOYOU.** MCCARTNEY and MANILOW have first names Paul and Barry, like the Chuckle Brothers, known for their catchphrase 'To me. To you.'

(c)   **DUNCAN.** SANTA and DECEMBER contain ANT and DEC, who portrayed PJ and Duncan.

193.

audIOTApe, motheRHOod, caPSIze, hOME GAme – each contains a Greek letter, now re-ordered into Greek alphabetical order.

194.

The title of this question is an anagram of Country names.

The answers are all anagrams of country names, with the corresponding countries being in alphabetical order:

(a) Regalia (Algeria)

(b) Chain (China)

(c) Denisonia (Indonesia)

(d) Rain (Iran)

(e) Serial (Israel)

(f) Laity (Italy)

(g) Liam (Mali)

(h) Reign (Niger)

(i) Pure (Peru)

(j) Angeles (Senegal)

(k) Rabies (Serbia)

(l) Yaris (Syria)

(m) Tango (Tonga)

195.

(a) **FIBONACCI**: the letters are replaced by A-1, B=1, C=2, D=3, E=5, F=8, G=13, H=21, I=34, J=55, K=89, L=144, M=233, N=377, O=610, ... (the Fibonacci sequence).

(b) **PASCAL**: the letters are replaced by A=1, B=3, C=6, D=10, E=15, F=21, G=28, H=36, I=45, J=55, K=66, L=78, M=91, N=105, O=120, P=136, Q=153, R=171, S=190, ...(the triangular numbers, which appear in Pascal's triangle).

(c) **Euclid's**: the elements' symbols are Eu, Cl, I, Ds respectively.

196.

Charmaine's lottery entry is:

Charmaine   11 31 32 38 42 45

Each letter in the name is replaced by its numerical value (using A=1, B=2, etc). The resulting 12-digit number is divided into 6 2-digit numbers. Each of these is replaced by its equivalent modulo 49 to give the lottery entry. Thus:

Anthony   → 114208151425 → 11/42/08/15/14/25 → 11 42 8 15 14 25

Charmaine → **381181319145** → **38/11/81/31/91/45** → **38 11 32 31 42 45**

Dorothy   → 415181520825 → 41/51/81/52/08/25 → 41 2 32 3 8 25

Geraldine → 751811249145 → 75/18/11/24/91/45 → 26 18 11 24 42 45

Justine   → 102119209145 → 10/21/19/20/91/45 → 10 21 19 20 42 45

Orville   → 151822912125 → 15/18/22/91/21/25 → 15 18 22 42 21 25

197.

All are properly written with an apostrophe-like punctuation mark. As is 'what's'! The marks are apostrophe, okina and ayn'.

Hallowe'en, Hawai'i, 'Oumuamua, Qur'an, Tian'anmen

198.

(a) The words to the left of the colon have a prime Scrabble score, the word to the right doesn't: AVOCADO (13), BALUSTRADE (13), CLOTHO (11), DATABASE (11), ENTENTE (7), FABULOUS (13), GEOGRAPHIC (19), HOMESICK (19), INQUISITOR (19), JACARANDA (19): KALAHARI (15)

(b) The words to the left of the colon have a prime word score (using A=1, B=2, etc), the word to the right doesn't: AVOCADO (61), BALUSTRADE (103), CLOTHO (73), DATABASE (53), ENTENTE (83), FABULOUS (97), GEOGRAPHIC (89), HOMESICK (83), INQUISITOR (151), JACARANDA (53): KLONDIKE (81)

(c) The words to the left of the colon have a prime text score (ABC=2, DEF=3, etc), the word to the right doesn't: AVOCADO (29), BALUSTRADE (47), CLOTHO (31), DATABASE (29), ENTENTE (37), FABULOUS (41), GEOGRAPHIC (43), HOMESICK (37), INQUISITOR (61), JACARANDA (31): KOSOVO (38)

199.

These are songs by Queen, King and Princess, so the table is completed by any single by Prince (e.g. Purple Rain). All such answers are right, though there are two that may be more relevant than others.

200.

The answers to the clues lead to people with alliterative names.

(a) The composer wrote the ENIGMA VARIATIONS. (Edward Elgar)

(b) The comms officer was Lt. UHURA. (Nichelle Nichols)

(c) The Tsar was known as the TERRIBLE. (Ivan IV)

(d) The actress wanted to be ALONE. (Greta Garbo)

(e) The punk innovator managed the SEX PISTOLS. (Malcolm McLaren)

(f) The stadium hosts the US OPEN final. (Arthur Ashe)

201.

(a) Blues Funeral	(f) Barry John
(b) Stoney Middleton	(g) Castle Bolton
(c) Middleton Stoney	(h) Funeral Blues
(d) Aye aye	(i) Aye-Aye
(e) Bolton Castle	(j) John Barry

Pairs: (a) with (h), (b) with (c), (d) with (i), (e) with (g), (f) with (j)

202.

    (a)   **5323.** In this cipher A=2, B=3, C=5, D=7, etc, i.e. primes. The question reads: WHAT IS THE ENCODING OF THREE POINT ONE FOUR ONE ETC? To which the answer is PI.

    (b)   **9872134.** In this cipher A=1, B=1, C=2, D=3, R=5, etc, i.e. the Fibonacci sequence. The question reads: AND SIMILARLY WHAT IS THE ENCODING OF ONE POINT SIX ONE EIGHT ZERO THREE ETC? To which the answer is PHI.

203.

ARCH, AY, GUST, EMBER. They are words that can be found at the ends of the names of months, March, May, August, September/ November/December.

204.

They are the signs of the zodiac in Dutch (RAM, STIER, TWEELINGEN, KREEFT, LEEUW, MAAGD, WEEGSCHAAL, SCHORPIOEN, BOOGSCHUTTER, STEENBOK, WATERMAN, VISSEN), English (ARIES, TAURUS, GEMINI, CANCER, LEO, VIRGO, LIBRA, SCORPIO, SAGITTARIUS, CAPRICORN, AQUARIUS, PISCES), French (BELIER, TAUREAU, GEMEAUX, CANCER, LION, VIERGE, BALANCE, SCORPION, SAGITTAIRE, CAPRICORNE, VERSEAU, POISSONS) and Swedish (VADUREN, OXEN, TVILLINGARNA, KRAFTAN, LEJONET, JUNGFRUN, VAGEN, SKORPIONEN, SKYTTEN, STENBOCKEN, VATTUMANNEN, FISKARNA). The list of 48 has been enciphered with the keyphrase ZODIAC IN DUTCH ENGLISH FRENCH (AND) SWEDISH and then sorted alphabetically; diacritics have been ignored.

Plain: ABCDEFGHIJKLMNOPQRSTUVWXYZ
Cipher: ZODIACNUTHEGLSFRWBJKMPQVXY

205.

Greece and Albania. You may have also found Albania and the disputed territory Kosovo.

206.

The groups correspond to the 6 quarks, Up, Down, Top, Bottom, Charm, Strange:

Words that can follow TOP (GEAR, GUN, MARKS)

Words that mean BOTTOM (BASE, FATHOM, LOWEST)

Words containing UP (CUPOLA, LUPINE, SUPPORT)

Words that can precede DOWN (BREAK, SHOW, TOUCH)

Words found (with gaps) in STRANGE (AGE, RAG, SAGE)

Words found inside CHARM (ARM, CHAR, HARM)

207.

(2,2,1,1,2,20). This represents twenty. Letters are represented by the first number in which they appear. The question reads: What numerals should stand for the letters in the answer to four times five?

208.

The pairs form the names of capital cities when pronounced together. Pairs are:

AL JEERS (ALGIERS), BAN JOULE (BANJUL), BAY ROUTE (BEIRUT), HAVE ANNA (HAVANA), CAR BULL (KABUL), CAR TOMB (KHARTOUM), LEA MAR (LIMA), LOW MAY (LOME), KEY TOE (QUITO), WORE SORE (WARSAW)

209.

lynx. Calculate the result of each formula. Each digit before the decimal point tells you a digit to extract from the number after the decimal point (0-up). This gives the integer associated with the formula. So π, for example, is at position 5 because the third digit (0-up) of .141592... is 5. And for 18: 2pi + 7e is 25.3111581... – and the digits in positions 2 and 5 give 18. For the formulae with variables in them, take the lowest positive integer value for the variable that gives the right solution: l=2, m=6, n=5, x=7, y=3, z=9.

Solve the final formula to get 210. $210 = l \times y \times n \times x$.

210.

In order they are:

0 1 1 1 (an OR gate)

0 1 1 0 (an XOR gate)

0 0 0 1 (an AND gate)

211.

Each is a number with the first letter changed:

SOUR, JIVE, FIX, NIGHT, PINE, PEN, SHIRTY, NIFTY

FOUR, FIVE, SIX, EIGHT, NINE, TEN, THIRTY, FIFTY

212.

(a) Each set of letters is a number advanced through the alphabet by another number, with which it is paired:

AVAR (= NINE + 13)	pairs with	CQRACNNW (= THIRTEEN + 9)
BPZMM (= THREE + 8)	pairs with	HLJKW (= EIGHT + 3)
JMZI (= FIVE + 4)	pairs with	KTZW (= FOUR + 5)
OKPWU VYQ (= MINUS TWO +2)	pairs with	RUM (= TWO −2)
UFO (= TEN + 1)	pairs with	YXO (= ONE + 10)

(b) The pairs are names of ciphers encoded using that cipher. Pairs are:

FDHVDU FLSKHU = CAESAR CIPHER

KTCGAL KRIKQTQRQTFD = SIMPLE SUBSTITUTION

LAYF PYRB = PLAYFAIR enciphered using Playfair with key PLAYFAIR

TSINR PTxAOIxNSOx = TRANSPOSITION

ASxUBE IIDNPONLTOSTOR = DOUBLE TRANSPOSITION

KJDYPFKAPUAQTB QROPRCRRRCJH = POLYALPHABETIC SUBSTITUTION

The simple substitution used the alphabet:

Plain: ABCDEFGHIJKLMNOPQRSTUVWXYZ
Cipher: SIMPLEUBTONACDFGHJKQRVWXYZ

The transposition was formed by reading down the columns in order:

TRAN
SPOS
ITIO
Nxxx

The double transposition was formed by reading off the columns in alphabetical order

DOUBL
ETRAN
SPOSI
TIONx

to get BASN DEST LNIx OTPI UROO, and then repeating

BASND
ESTLN
IxOTP
IUROO

to get ASxU BEII DNPO NLTO STOR.

The polyalphabetic substitution used the following alphabets:

Plain: ABCDEFGHIJKLMNOPQRSTUVWXYZ
Cipher 1: polyahbeticdfgjkmnqrsuvwxz
Cipher 2: SUBTIONACDEFGHJKLMPQRVWXYZ

POLYALPHABETIC SUBSTITUTION

kJdYpFkApUaQtB qRoPrCrRrCjH

213.

Purple. The colours are those of the Mr Men characters with those names. Mr Impossible is Purple.

214.

one hunDreD: 1000 = 10 × 100

one hunDreD anD sIXty eIght: 1512 = 9 × 168

two hunDreD: 1000 = 5 × 200

three hunDreD anD eLeVen: 1555 = 5 × 311

fIVe hunDreD anD two: 1506 = 3 × 502

seVen hunDreD anD fIfty three: 1506 = 2 × 753

215.

(a) **X.** Upper-case letters common between Greek and Latin alphabets.

(b) **T.** Upper-case letters common between Russian Cyrillic and Latin alphabets.

216.

(a) Numbers written as words, with each letter replaced by the first number in which that letter appears. Next in sequence is EIGHT = **15832**

(b) Maximum cost in pounds of landing on successive Monopoly spaces, starting at Old Kent Road:

250, 2660, 450, **200**, 200, 550, 2660, **550**, 600, 50, 750...

In the seventh position, £2660 is the cost of the 'Street Repairs' Chance card if you have 4 houses on 8 of the properties and a hotel on 12 of the remaining properties (a set has only 32 houses and 12 hotels), for a total cost of 32 x £40 + 12 x £115 = £2660. In the second position, the same value is derived by drawing 'Pay £10 fine or take a Chance' from Community Chest, opting to draw a Chance card and getting 'Street Repairs'.

(c) Superimposed semaphore for 'To be or not to be that is the' = **question** =

217.

> **Y.** Extracting the ordinal from English monarchs, William I, wIlliam II, Henry I, Stephen (I), hEnry II, Richard I, John (I), heNry III, Edward I, eDward II, edWard III, rIchard II, henRy IV, henrY V. The sequence ends here, because next in succession was Henry VI, but Henry only has 5 letters.

218.

> liechTENstein, JACKsonville, QUEENsland, united KINGdom, mACEdonia all contain playing cards. Since Aces can sometimes be low, you may also put Macedonia at the start of the list.

219.

> (a) **125.** The sequence consists of the lowest numbers containing 3, 4, 5, etc, different letters.
>
> ONE, THRE, EIGHT, THIREN, FOURTEN, TWENYFIV, TWENYFOUR, SEVNTYFOUR, ONEHUDRAFIV, ONEHUDRATWLV, ONEHUDRATWYSV, ONEHUDRATWYFIV

> (b) **28, 24000, 1000300, 108, 107.** If the sequence is written out in a column then the numbers ZERO, ONE, TWO, THREE, FOUR, … etc, read down the leading diagonal. In each case the numbers are the smallest possible that could be in the list:
>
> ZERO
> ZERO
> ZERO
> ZERO
> FOUROCTILLION
> ELEVEN
> THIRTEEN
> SEVENTYTWO
> SEVENTYTWO
> SEVENTYTWO

TWENTYEIGH<u>T</u>
TWENTYFOURT<u>H</u>OUSAND
ONEMILLIONTH<u>R</u>EEHUNDRED
ONEHUNDREDAND<u>E</u>IGHT
ONEHUNDREDANDS<u>E</u>VEN
ONEHUNDREDANDF<u>I</u>FTEEN
THREEHUNDREDANDF<u>O</u>UR
THREEHUNDREDANDF<u>O</u>UR
THREEHUNDREDANDFO<u>U</u>R

220.

(a) Using A=1, B=2,..., gives

CAP, DUE, FIX, GEM, PRO (or **AFRO**), **FROM**

(b) Using the periodic table H=1, He=2, Li=3, ... gives:

GAS, MOP, CUBE, REAL, SHOP, CARPAL

221.

The answers to the clues give numbers which if typed on a phone pad keyboard give the letters in enigma.

(a) The American beer is ROLLING ROCK. (33 appears on every bottle.)

(b) The Highway ends in SANTA MONICA. (Route 66.)

(c) Internet pagers use the SIMPLE NETWORK PAGING PROTOCOL. (SNPP uses port 444.)

(d) The album is by ONE DIRECTION. (Album is called Four.)

(e) Danger Man was followed by THE PRISONER. (Number Six in the TV series.)

(f) The Bond villain was played by ADOLFO CELI. (Emilio Largo is No. 2 in Spectre.)

222.

    (a) **BRIDGE OVER TROUBLED WATER.** Converting the numbers into base 36 gives BRIDGE/AEWRT.

    (b) **TAKE FIVE.** 5 has been subtracted/taken (cyclically) from each letter.

    (c) **THREE TIMES A LADY. LADY D'ARBANVILLE, LADY ELEANOR** and **LADY MADONNA** are songs by Cat Stevens, Lindisfarne and The Beatles respectively.

    (d) **WITHOUT YOU.** The phrase 'A NUMBER ONE HIT SINGLE' is without U.

    (e) **THE TIMES THEY ARE A CHANGIN'.** THREE MAY HESITATE is an anagram of THE TIMES THEY ARE A.

    (f) **LETTER FROM AMERICA.** The letters are those of AMERICA.

    (g) **MESSAGE IN A BOTTLE.** The letters of MESSAGE are embedded within METHUSELAH.

223.

U. These are the last letters of ALPHA, BETA, GAMMA, etc. U is the last letter of MU. The title clues the idiom 'It's all Greek to me'.

224.

Leopards in the barrows = Borrowed Latin phrases

These are all anagrams.

    (a) In loco parentis

    (b) Semper fidelis

    (c) Carpe diem

    (d) Sic transit gloria mundi

    (e) Modus operandi

    (f) Persona non grata

    (g) Habeas corpus

    (h) Per capita

    (i) Non sequitur

    (j) Magna carta

    (k) Ipso facto

    (l) In vino veritas

225.

A secret never to be told.

```
   T
   H           I       S O                 D
   S I     O Y N     S E N B     P X E I
 F I R S T R E D F I V E E P S I L O N
 O X D E E A L I O X E     T I I     T T U
 U T     V N N L G U     N     A         A A
 R H     E T G O O R     M E R C U R Y
 T       N H E W       E A R T H L A N D O
 H       T J U P I T E R S O T E R
         H       V E N U S       J O H N
```

The pieces are created by linking the members of each set together, for example:

```
   T
   H
  SI
FIRST
OXDEE
UT  VN
RH  ET
T   NH
H   T
    H
```

The 6 pieces then fit into the grid. The grey squares, read backwards, give SEVENTH MAGPIE.

226.

NEED, LIBRA, ROMANCE, YAM. They are anagrams of British Prime Ministers Eden, Blair, Cameron and May, in order of their terms of office.

227.

COPERNICIUM, CURIUM, MANGANESE, MOSCOVIUM, NIOBIUM, ZINC. The top shelf contains those elements that have a 2-letter symbol that can be formed using the letters on the top row of a typewriter keyboard; the middle shelf contains those similarly corresponding to the second row; the lower shelf similarly corresponds to the third row.

228.

HAZARD is the odd word out. The other words contain letters inside synonyms for 'obvious'. BASALTIC, COCHLEAR, ECSTASY, PLANTAIN

229.

Words made up entirely of Roman numerals. The corresponding numerical values are the sum of the individual letter values, ignoring the ordering.

(a) Livid = 557

(b) Cilic = 252

(c) Dili = 552

(d) Vidic = 607

(e) Vivimi = 1013

(f) Civic = 207

(g) Limici = 1153

(h) Dimili = 1553

(i) Diili = 553

(j) Illici = 203

230.

OLD. One letter in each word can be changed to give the question WHAT IS THE ODD WORD OUT? The letter changed is the first, except for OLD where it is the second – thus OLD is the answer to this question.

231.

Each word is a number in a foreign language.

DUE (2 Italian), PUMP (5 Welsh), HAT (6 Hungarian), ELF (11 German), DOZE (12 Portuguese), CENT (100 French)

ELF (German) is ONCE (Spanish), both equal to 11.

232.

The function takes inputs from 0 - 31 and produces outputs in the same range. We can represent this on the following diagram, which shows what happens when we apply the function repeatedly (the 'cycle structure'):

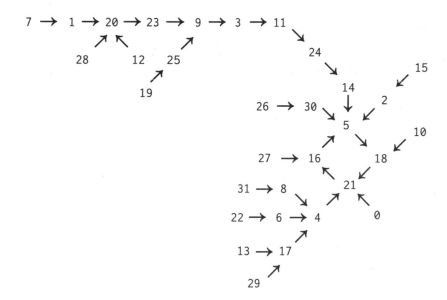

Using A=1, B=2, ..., Z=26 then leads to the diagram below. No matter where we start, we always end up cycling through four letters spelling out my destination, PERU. And at the top left we find that I'm flying from GATWICK.

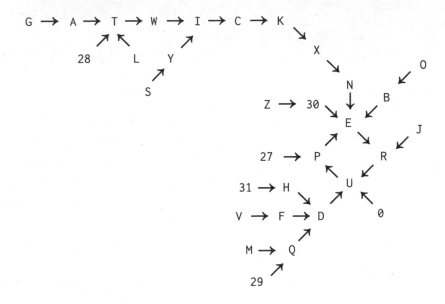

233.

Any of them. The words appear in the titles of the Harry Potter books:

philosophers sTONe, cHAMber of secrets, prisoner of azkaBAN, gobLET of fire, order of the pHOEnix, half-bLOOd prince – dEAThly hALLOWs

234.

Actors and films they were in:

Alan Alda	=	Bridge of Spies, Crimes and Misdemeanors, Same Time Next Year
Ben Barnes	=	Dorian Gray, Killing Bono, Seventh Son
Claudette Colbert	=	Cleopatra, It Happened One Night, Imitation of Life
Doris Day	=	Calamity Jane, Pillow Talk, That Touch of Mink

Emilio Estevez	=	The Breakfast Club, The Mighty Ducks, Young Guns
Frances Fisher	=	House of Sand and Fog, Titanic, Unforgiven
Greta Garbo	=	As You Desire Me, Ninotchka, Wild Orchids
Holly Hunter	=	Broadcast News, Raising Arizona, The Piano
Ian Ignacio	=	Ang Bagong Dugo, Showdown in Manila
Jennifer Jones	=	Cluny Brown, Love is a Many-Splendored Thing, Love Letters
Keira Knightley	=	Bend it Like Beckham, Pride and Prejudice, The Imitation Game
Lindsay Lohan	=	Freaky Friday, Herbie Fully Loaded, Mean Girls
Matthew McConaughey	=	Dallas Buyers Club, Free State of Jones, Interstellar
Nanette Newman	=	The Raging Moon, The Stepford Wives, The Whisperers
Ozzy Osbourne	=	Austin Powers in Goldmember, Ghostbusters, Little Nicky
Piper Perabo	=	Because I Said So, Coyote Ugly, Looper
Quentin Queensbury	=	The Blinds
Robert Redford	=	A Walk in the Woods, Indecent Proposal, Out of Africa
Sylvester Stallone	=	Oscar, Rambo, Rocky
Thelma Todd	=	Horse Feathers, Monkey Business, You Made Me Love You
Ursula Ulrich	=	Hurrah! I'm a Papa, Many Lies
Vince Vaughn	=	Couples Retreat, The Internship, Wedding Crashers
Wil Wheaton	=	Pie in the Sky, Stand By Me, The Last Starfighter
Xixi Xiao	=	Ru Ci Die Niang

Yves Yan = L'âme du Tigre, Mon Héros, Virtual Revolution

Zhang Ziyi = Crouching Tiger Hidden Dragon,
(or Ziyi Zhang) Sophie's Revenge, Zu Warriors

235.

SPY.

236.

(a) **Look backward**

(b) **IGNORE PRIME LETTERS**: ImaGiNe OREs PuRIM bElLE ToTERS

(c) **PAIR THEN ORDER.** The words can be paired to create a letter from the NATO phonetic alphabet:

S-PA PA-MPA	P
UNRE-AL FA-CADE	A
S-IN DIA-BLO	I
PALIND-ROME O-TTO	R
TI-TAN GO-D	T
W-HO TEL-EPORTER	H
CHOK-E CHO-KE	E
USTI-NOV EMBER-S	N
TYP-OS CAR-LESS	O
F-ROM EO-M	R
MO-DEL TA-NGO	D
QUEB-EC HO-TEL	E
SUPERHE-RO MEO-W	R

237.

e.g. 20113 and 10293, and variations on this:

TWENTY THOUSAND ONE HUNDRED AND THIRTEEN

TEN THOUSAND TWO HUNDRED AND NINETY THREE

238.

NEWER SWELL, SUDDENNESS, DULL SENSE, NEWS LURED. Interpreting every occurrence of UDLR and NSWE as an instruction to move Up, Down, Left, Right, North, South, West or East (and equating the two systems in the obvious way), we can draw the following shapes:

DULL SENSE   NEWER SWELL   NEWS LURED   SUDDENNESS

Rotating the shapes reveals the digits 8, 0, 9 and 5, which we place in numerical order. The 9 can also be read as 6, which would make the answer NEWER SWELL, SUDDENNESS, NEWS LURED, DULL SENSE.

239.

(a) **Mercury** Rev

(b) **Venus** Williams (or Michael **Venus**)

(c) **Earth** Song

(d) Veronica **Mars**

(e) Sally **Jupiter** (or Laurie **Jupiter**)

(f) Sega **Saturn**

(g) **Uranus** Hill/Quarter

(h) King **Neptune**

240.

The missing words can be preceded by the letters in Enigma.

(a) Novocaine is a NUMBER – **E** number.

(b) The abbreviation for synchronization is SYNC – **N***SYNC.

(c) AIBO is a ROBOT – **I**, Robot.

(d) The rate of change of momentum is proportional to FORCE – **G**-Force.

(e) The square is TWENTY FIVE – **M**25.

(f) The tributary of the Tyne is the TEAM – **A**-Team.

241.

14.7. 14 July is Bastille Day, the national day of France.

242.

MUSIC. Each bout is a game of ROCK, PAPER, SCISSORS, represented by synonyms.

SWING *v* ESSAY	ESSAY		
CUTS *v* MAIL	CUTS	CUTS	
			MUSIC
DEED *v* HOLD	HOLD	MUSIC	
MUSIC *v* SNIPS	MUSIC		
			MUSIC
RAG *v* TOR	RAG		
SHEARS *v* THESIS	SHEARS	SHEARS	
			SHEARS
MIRROR *v* COBBLE	MIRROR	MIRROR	
ZIGZAGS *v* DISTURB	DISTURB		

243.

UG. Snake Eyes. The table represents the initials of the names of throws in Craps, advanced through the alphabet by their score.

DG=3+AD (Ace Deuce)

IJ=4+EF (Easy Four)

LJ=4+HF (Hard Four)

KK=5+FF (Fever Five)

KY=6+ES (Easy Six)

NY=6+HS (Hard Six)

U=7+N (Natural)

MM=8+EE (Easy Eight)

PM=8+HE (Hard Eight)

W=9+N (Nine)

OD=10+ET (Easy Ten)

RD=10+HT (Hard Ten)

J=11+Y (Yo-leven)

N=12+B (Boxcars)

244.

(a)   ISBNs of consecutive Ordnance Survey Explorer maps:

9780319243664 = Map 173: London North, The City, West End, Enfield, Ealing, Harrow & Watford

9780319243671 = Map 174: Epping Forest & Lee Valley

9780319243688 = Map 175: Southend-on-Sea & Basildon

9780319243695 = Map 176: Blackwater Estuary

9780319243701 = Map 177: Carmarthen & Kidwelly

9780319243718 = Map 178: Llanelli & Ammanford/Rhydaman

9780319243725 = Map 179: Gloucester, Cheltenham & Stroud

(b) **PACED (4).** Shift around the alphabet by the given number of places, then anagram to get numbers:

BAR	shift by 13	ONE	anagram	ONE
LOG	shift by 8	TWO	anagram	TWO
PANDA	shift by 4	TERHE	anagram	THREE
DRAG	shift by 14	RFOU	anagram	FOUR
BEAR	shift by 4	FIEV	anagram	FIVE
TOE	shift by 4	XSI	anagram	SIX
FRIAR	shift by 13	SEVNE	anagram	SEVEN
PACED	shift by 4	TEGIH	anagram	EIGHT

(c) Lengths of words of the digits of π:

3.14159265358979323846 ...

5.34344334545454535543 ...

245.

BOB ROSS is the odd one out because all the letters in his name have curved edges, whereas all the other names consist of letters made of only straight edges.

246.

One possible answer is:

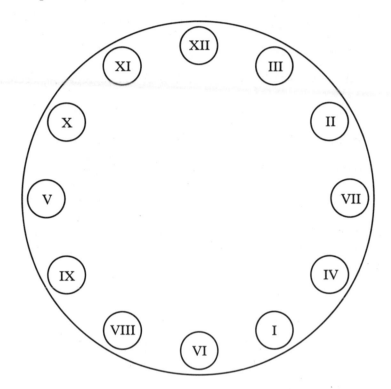

247.

Myrrhy Christmas. The 3-letter word used as the keystream is OIL. The words are all oils:

<u>M</u>ustard <u>Y</u>arrow <u>R</u>ose <u>R</u>osehip <u>H</u>enna <u>Y</u>arrow

<u>C</u>oconut <u>H</u>elichrysum <u>R</u>osemary <u>I</u>pecac <u>S</u>pruce <u>T</u>arragon <u>M</u>ugwort <u>A</u>nise <u>S</u>assafras

This makes a 'seasonal' greeting!

248.

> **Ventriloquists and their dummies**: Darci Lynne & Petunia, Keith Harris & Orville, Ray Alan & Lord Charles, Roger de Courcey & Nookie Bear, Shari Lewis & Lamb Chop

249.

> Surnames of characters whose first names provide film titles:
> **Alfie** Elkins (1966), **Sabrina** Fairchild (1954), **Hanna** Heller (2011), **Dave** Kovic (1993), **Juno** MacGuff (2007), **Amélie** Poulain (2001), **Matilda** Wormwood (1996)

250.

> Using 2, 4, 6, 8 to represent ↓, ←, →, ↑ respectively, as on the numeric keypad of a keyboard, the sequence traces out the following shape, starting at the bottom left:

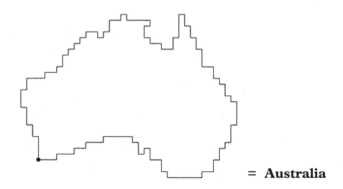

= **Australia**

251.

> This is a circuit made up of binary logic gates. Labelling the input bits at the bottom from left to right as (a, b, c, d, e), the function is:
>
> $$c(\bar{d}(a \lor b) \oplus d\,(b \oplus \bar{e})) \oplus \bar{d}e$$
>
> For $(a, b, c, d, e) = (0,0,0,0,0), (0,0,0,0,1), (0,0,0,1,0), \ldots (1,1,1,1,1)$, this gives the following 32 binary output values:
>
> 01000110010010010100101001001001 = 46494A49 (hex)
>
> Treating this as four ASCII bytes, it spells **FIJI**.

252.

The question reads 'What is next in the sequence?' In each series and answer only the consonants are given.

(a) **MCRN.** Mu, Nu, Xi, oMiCRoN. Greek letters in order

(b) **MCRN.** CHiRaC, SaRKoZY, HoLLaNDe, MaCRoN. French Presidents in order

(c) **MCRN.** aNGSTRoM, NaNoMeTRe, TeN NaNoMeTReS, oNe HuNDReD NaNoMeTReS, MiCRoN. Each measure is 10 times the previous one

Each answer is MCRN, as are the remaining answers:

WHiCH SoNG?   **MaCaReNa**

WHiCH PeNGuiN?   **MaCaRoNi**

253.

All of them. The words in the question can have a letter added to form an animal:

wasP, bEar, sNail, tiGer, loUse, plaIce

These letters spell out PENGUI, so the final word needs to be one to which an N can be added to form an animal. They all have this property: aNt, heroN, raveN, sNake.

Entrance Tests

A note about the entrance-test questions

These entrance-test questions were found a few years ago in a bundle of papers at the bottom of a filing cabinet. Unfortunately, the papers did not include the solutions to the questions and so we have had to reconstruct them. In some cases, the setters have left deliberate ambiguities in the question, with the solver required to make a best guess as to what the right answer should be. We hope that we have interpreted these ambiguities correctly in our solutions. Furthermore, it is possible that some of these questions have more than one solution. We make no guarantee that our answer will always match up with yours (although we expect it will, in most cases).

254. 1968 Entrance Test

Problem 1:

XOXOOXXOOOXXXOXXOXOXOOOOX (i.e. the occupied seats are: A, C, F, G, K, L, M, O, P, R, T, Y)

The incorrect statements are **10** and **18**.

(Note that statements 17 and 18 are inconsistent, as is the group 6,7,8 and 10, and so the remaining statements must be correct – proceed from there).

Problem 2:

12 members were present, as shown – the numbers indicate where the members who made the statements were sitting.

```
   6   2    5  4      3    1
OXXXOXXOOOXOXXOXOOXXOOXOO
```

255. **1975 Entrance Test I**

1) TIDELESS LTD ANNOUNCED YESTERDAY A NEW
BIOLOGICAL PRODUCT CALLED BANG WHICH IS
CLAIMED TO BE CAPABLE OF REMOVING SPOTS
FROM LEOPARD SKINS. BANG WILL BE MARKETED
EXPERIMENTALLY OVER THE NEXT 6 MONTHS IN
THE WEST CORNWALL AREA.

2) IT IS REPORTED THAT MR. J. BUCKFAST WILL
ANNOUNCE HIS RESIGNATION AS CHAIRMAN
OF THE DETERGENT GIANT DREFTWOOD AND
FLASHBACK LATER THIS WEEK. HIS RESIGNATION
HAS BEEN EXPECTED EVER SINCE DREFTWOOD AND
FLASHBACK DECLARED BEFORE-TAX PROFITS FOR
1973/4 OF £73.31.

In the first message all letters after L in the alphabet have been
replaced with L; in the second all those after J have been replaced with
J. Some of the words – especially the name in part 2 – are best guesses
and cannot be determined with any certainty.

256. **1975 Entrance Test II**

48061 53829 76351 63140

The first 5-digit number in row n is the last 5-digit number in row
n on the previous page, but read backwards and then rotated by n-1
places to the left (so, for example, to get the first value for page 2, row
2, take the final value for page 1, row 2 (i.e. 15679), reverse it (to get
97651) and then rotate it by 1 place to the left to get 76519).

For each subsequent 5-digit number, the number in the same row,
previous column on the previous page is taken, but a digit substitution
cipher has been applied: $0 \to 6 \to 2 \to 4 \to 1 \to 9 \to 3 \to 8 \to 5 \to 7$
$\to 0$ (i.e. each 0 is replaced by a 6, each 6 by a 2, and so on). So, for
example, 98402 in page 1, row 2, column 1 becomes 35164 in page 2,
row 2, column 2.

257. 1976 Entrance Test I

(a) The boys invented **212, 415, 560, 632, 737, 884, 999**

(b) **121**

(c) **026, 096, 166** (i.e. derived from 602, 609, 616)

The numbers are multiples of 7 with the first digit shifted to the end (note the cheeky clue of calling the island Sevensey). For example, 7 × 43 = 301 and therefore 013 is a legitimate number.

The final digit represents the parish where a car is registered. Given that 3 is the tail digit assigned to St Aimée cars and 5 is the tail digit for St Pancras, the following seems a plausible assignment of tail digits for the parishes, given the order that they are provided in the question: 1–2 are assigned to St Margery, 3 to St Aimée, 4 to St Dottie and 5 to St Pancras. Also, given that the Chair of the St Pancras parish council has number 045 (which is derived from 504, i.e. the smallest multiple of 7 after 500), it seems likely that the Deputy Chair would therefore get the second smallest number in the relevant parish's sequence.

258. 1976 Entrance Test II

After printing a lower-case m, the wheel skips one position backwards so that all subsequent characters need to be shifted forwards alphabetically. Thus the tail of the message should read:

> 'I was given one for Christmas. It is most unreliable. What has gone wrong now please?'

259. 1977 Entrance Test

Firstday, Houndday, Dragonday, Snakeday, Ramday, Catday, Beeday, Ibisday, Orcday, Assday, Lastday

It seems reasonable to assume that the week begins with Firstday and ends with Lastday, in which case this is the only order consistent with the kings' dates.

260. 1980 Entrance Test

When the pages are numbered correctly and sorted, they are labelled 'from' and 'to' as follows:

Page no.	From	To
1	2,3,5	3,5
2	3	1,4
3	1	1,2
4	2,7,10	6,7,9
5	1	1
6	4	8
7	4	4
8	6,12,14	11,13,14
9	4	10
10	9	4
11	8	12,14
12	11	8
13	8	15
14	8,11	8
15	13,17,19	16,17,18
16	15	19
17	15	15
18	15,20	–
19	16	15,20
20	19	18

261. 1981 Entrance Test

There are multiple solutions to this but the comment 'there is a uniquely convincing solution' can only mean that the following must be the intended solution:

Y E S C O M M A T

H I S I S O B V I

O U S L Y T H E R

I G H T A N S W E

R C O L O N T H E

Q U I C K B R O W

N F O X J U M P S

O V E R T H E L A

Z Y D O G S T O P

262. 1984 Entrance Test

The grid, with plausible answers entered for the unknown across clues, is as follows:

D	E	C	A	M	P		C	O	M	I	C	S
	N		B		R	N	R		A		A	
C	A	N	D	L	E		E	X	C	U	S	E
	C		O		S		D				T	
A	T	O	M		S	N	O	W	D	R	O	P
			I		U			E		F		
T	R	A	N	S	P	O	R	T	C	A	F	E
	E		A				A		E			
C	A	U	L	D	R	O	N		P	I	L	E
	L			U		S		T		O		
T	I	G	H	T	S		A	N	I	M	U	S
	S		O		T	U	C		O		S	
S	T	U	B	B	Y		K	E	N	N	E	L

The key is to realize that 'Café' isn't a complete solution, but part of 'Transport Café'. (There are other possible answers for several of the across clues.)

263. 1985 Entrance Test

The keyword is **VENTRILOQUY**: write it in a 5x5 square and read down the columns.

V E N T R

I L O Q U

Y A B C D

F G H K M

P S W X Z

264. **1986 Entrance Test**

WORSHIPFUL, with the letters representing 1, 2, 3, 4, 5, 6, 7, 8, 9 and 0 respectively.

The misread prices are £AAL (actually £RRL), and £FW (presumably £WF).

A Right Royal Puzzle

The answers to the seven questions are all names of British monarchs. In particular, the answers take the form of the monarch's regnal name (e.g. Elizabeth) and the monarch's regnal number (which is 2, in the case of Her Majesty the Queen).

265.   A right royal puzzle: question 1

The items listed are all anagrams of ingredients in the cocktail known as a 'Bloody Mary'.

CLAY KESTREL	CELERY STALK
CRYSTAL EEL	CELERY SALT
CUBA SEACOAST	TABASCO SAUCE
CUBIC SEE	ICE CUBES
JAM TOO, CUTIE	TOMATO JUICE
JOULE MINCE	LEMON JUICE
LEWD GENOME	LEMON WEDGE
P. E. PREP	PEPPER
REOCCURS, WEATHERISES	WORCESTERSHIRE SAUCE
D. V. OAK	VODKA

'Bloody Mary' was the nickname of **Mary I**, which is the answer to this question.

266. A right royal puzzle: question 2

The people listed in this question all had different titles:

(1) Thomas I

(2) Edward II

(3) ???

(4) 4th Duke of Gordon

(5) James V

If you spotted all the train-related clues in the question, you might have then guessed that this question has to do with the locomotives from the Thomas the Tank Engine stories. Thomas the Tank Engine had a 1 painted on his side, Edward was the number 2 engine, Gordon was number 4, and James was number 5. Number 3 engine was Henry, and so the answer to this question is **Henry III**.

267. A right royal puzzle: question 3

Here are the original names with their celebrity pseudonymizer counterparts:

JOHNS OGOGO	ANTHONY HOPKINS
ULYSSES BONNEVILLE	HUGH GRANT
NATHANIEL PARGETTER	NIGEL HAWTHORNE
MERRILEE CHAUCER	GEOFFREY RUSH
BOONE REDFORD	ROBERT CARLYLE

These actors won the BAFTA Award for Best Actor in a Leading Role in the years 1993, 1994, 1995, 1996 and 1997.

Nigel Hawthorne won for his role as George III in the film The Madness of King George, and so the answer is **George III**.

268. A right royal puzzle: question 4

(a) The first letter of each of these four strings of letters is H, J, L and N. J is two above H in the alphabet, L is two above J, and so on, and so the next letter in the sequence is P. The same rule applies to the letters in the strings, e.g. the second letters are J, L, N and P, and so on. If you work out what the next letter is for each position then you get the word PRINCE, which is therefore the next string in the sequence.

(b) These are the chemical symbols for Fluorine, Sulfur, Manganese and Krypton. Their atomic numbers are, respectively, 9, 16, 25 and 36, which are successive square numbers. The next square number is 49 and the element for which that is the atomic number is Indium. The chemical symbol for Indium is In.

(c) These are cards from the Major Arcana suit in a Tarot deck of cards. Their numbers are respectively 1, 2, 4 and 8, which are the powers of 2 in order. The next power of 2 is 16 and the card that corresponds to that number is THE TOWER.

If you put the three answers together you get: PRINCE IN THE TOWER. Hopefully by now you will have worked out the theme that the answers to these questions are all historical monarchs and so this leads to the answer to this question being **Edward V**.

269. A right royal puzzle: question 5

The headlines in this puzzle have all been encrypted using a simple substitution method:

Plain: KINGCHARLESTWOBDFJMPQUVXYZ
Encrypted: ABCDEFGHIJKLMNOPQRSTUVWXYZ

by which we mean that K encrypts to A, I encrypts to B, and so on. These are the headlines that you should have found:

GENERAL STRIKE CAUSES CHAOS IN BRITAIN

EDWARD VIII ABDICATES – GEORGE VI IS NEW KING

TRYGVE LIE IS APPOINTED FIRST SECRETARY GENERAL OF UN

SWITZERLAND WINS THE FIRST EUROVISION SONG CONTEST

ENGLAND DEFEAT WEST GERMANY IN WORLD CUP FINAL

DROUGHT CONTINUES IN BRITAIN'S HOTTEST SUMMER

JEAN-CLAUDE DUVALIER IS OVERTHROWN AS PRESIDENT OF HAITI

BREAKTHROUGH IN GENETICS: DOLLY THE SHEEP IS CLONED

PLUTO LOSES ITS STATUS AS A PLANET

QUEEN ELIZABETH CELEBRATES NINETIETH BIRTHDAY

These headlines are taken from events that occurred in 1926 (the year of the Queen's birth), 1936, 1946, and so on, up to 2016.

The encrypted headlines were signed off as 'ABCDEFGHIJKLMN' which decrypts to KINGCHARLESTWO and so the answer to this question is **Charles II**.

270. A right royal puzzle: question 6

E	R	E	C	O	G	N	I	S	I	N	G	S	E	R	V	I
C		L		N		E		U		I		I		I		C
N	O	I	S	E	L	E	S	S		N	I	G	H	T	I	E
I		Z		R		D		T		E		N		U		A
S	E	A	F	O	W	L		A	S	T	R	O	N	A	U	T
M		B		U		E		I		Y		R		L		H
O	V	E	R	S	T	R	U	N	G		F	I	A	S	C	O
D		T								N						M
G	E	H	R	I	G		1	9	0	9		A	R	O	S	E
N				N										V		A
I	N	S	E	C	T		P	O	S	T	M	O	D	E	R	N
K		E		L		P		C		A		R		R		D
D	I	A	M	E	T	R	I	C		C	H	E	L	S	E	A
E		S		M		I		I		T		G		L		B
T	O	I	L	E	T	S		P	R	I	V	A	T	E	E	R
I		D		N		O		U		L		N		E		O
N	U	E	H	T	G	N	I	T	C	E	T	O	R	P	D	A

'Recognising service at home and abroad protecting the United Kingdom since 1909' appears on the plaque in Westminster Abbey honouring the Intelligence Services: MI5, MI6 and GCHQ.

Across
9. D
10. A
11. V
12. I
13. D
14. T
15. H
17. E
19. S
21. E
26. C
27. O
28. N
29. D

Down
2. K
3. I
4. N
5. G
6. O
7. F
8. S
16. C
18. O
20. T
22. L
23. A
24. N
25. D

The letter from each clue read in order yields:

DAVID THE SECOND, KING OF SCOTLAND

and so the answer to this question is **David II**.

271. A right royal puzzle: question 7

Stage 1: There are 36 plays in the First Folio:

The Tempest	The Two Gentlemen of Verona	The Merry Wives of Windsor
Measure for Measure	The Comedy of Errors	Much Ado About Nothing
Love's Labours Lost	A Midsummer Night's Dream	The Merchant of Venice
As You Like It	The Taming of the Shrew	All's Well That Ends Well
Twelfth Night	The Winter's Tale	King John
Richard II	Henry IV, Part 1	Henry IV, Part 2
Henry V	Henry VI, Part 1	Henry VI, Part 2
Henry VI, Part 3	Richard III	Henry VIII
Troilus and Cressida	Coriolanus	Titus Andronicus
Romeo and Juliet	Timon of Athens	Julius Caesar
Macbeth	Hamlet	King Lear
Othello	Antony and Cleopatra	Cymbeline

Stage 2: The letter she said was T. Removing titles starting with T leaves 25:

Measure for Measure	Much Ado About Nothing	Love's Labours Lost
A Midsummer Night's Dream	As You Like It	All's Well That Ends Well
King John	Richard II	Henry IV, Part 1
Henry IV, Part 2	Henry V	Henry VI, Part 1
Henry VI, Part 2	Henry VI, Part 3	Richard III
Henry VIII	Coriolanus	Romeo and Juliet
Julius Caesar	Macbeth	Hamlet
King Lear	Othello	Antony and Cleopatra
Cymbeline		

Stage 3: My friend's name begins with F. Removing titles containing a word beginning with a letter earlier in the alphabet than F leaves 16:

Measure for Measure	Love's Labours Lost	King John
Richard II	Henry IV, Part 1	Henry IV, Part 2
Henry V	Henry VI, Part 1	Henry VI, Part 2
Henry VI, Part 3	Richard III	Henry VIII
Macbeth	Hamlet	King Lear
Othello		

Stage 4: She doesn't like the word Henry. Removing these leaves 9:

Measure for Measure	Love's Labours Lost	King John
Richard II	Richard III	Macbeth
Hamlet	King Lear	Othello

Stage 5: Removing those which end in a consonant leaves 4:

Measure for Measure	Richard II	Richard III
Othello		

Stage 6: Removing those which contain an odd number of letters leaves just her favourite play:

Richard III

Solving the last part …

To summarize, here are the answers to the seven questions:

(1) Mary I
(2) Henry III
(3) George III
(4) Edward V
(5) Charles II
(6) David II
(7) Richard III

We asked you to convert each answer to a letter using the same process each time. The secret to this is to take the letter of the regnal name in each case which corresponds to the associated regnal number. So, Mary I becomes 'M', Henry III becomes 'N', and so on.

Here are the letters for all the answers:

Mary I → M
Henry III → N
George III → O
Edward V → R
Charles II → H
David II → A
Richard III → C

Take these letters and unscramble them to get the final answer to the whole set of puzzles: **MONARCH**

Tiebreaker Puzzles

272. Tiebreaker: question 1

No one answered this question, so we will leave it as a challenge!

273. Tiebreaker: question 2

The penguin message refers to 'endpapers', and implies that the next stage of the competition can be found there. However, as the book is a paperback it doesn't have endpapers, as these are found only in hardback books, where they join the covers of the book to the front/back pages. The next competition stage should correctly have been described as being on the inside front cover.

274. Tiebreaker: question 3

No one answered this question, so we will leave it as a challenge!

275. Tiebreaker: question 4

The first letters of each paragraph spell FLAG.

276. Tiebreaker: question 5

Page 141. The substitution example text is 'What might you turn to when you get stuck on a problem' and this has been enciphered with the keyword SEMAPHORE. Solving the inside back cover message required you to 'turn' semaphore characters.

Nine puzzles

277. Tiebreaker: question 6

(a) DOLLAR, FANFARE, PREMIUM, RELATION, TIRESOME, DORMITORY

1-7 = DO,RE,MI,FA,SO,LA,TI

(b) STURDY, ATHEIST, STARDOM, THIRSTY, THUNDER, STANDARD

1-4 = ST,ND,RD,TH (1st, 2nd, 3rd, 4th)

(c) BANJO, DECODE, IMPACT, SCAMANDER, AMALGAMATE, MAIDENHEAD

1-9 = DE,PA,NJ,GA,CT,MA,MD,SC,NH

(First 9 states to join the Union)

(d) COMANCHE, HOBGOBLIN, AFFLICTION, HENCEFORTH, LIBERALISM, BEACHCOMBER

1-9 = H,He,Li,Be,B,C,N,O,F

(First 9 elements)

(e) SWINE, UNWISE, WINNER, ASSESSES, NEWSWOMEN, SENSELESSNESS

1-8 = N,NE,E,SE,S,SW,W,NW

(f) PUP, FAFF, PUMP, COMFY, SPIFFY, FOPPISH

1-6 = pp,p,mp,mf,f,ff (musical dynamics)

(g) SPIV, IMPLY, INDEX, CLEVER

1-7 = I,V,X,L,C,D,M (Roman numerals)

(h) BESIDES, FORTIFY, TIMEWORN, AESTHETIC, FORBIDDEN, WORDSMITH

1-9 = ID,EN,TI,FY,TH,ES,EW,OR,DS

278.　Tiebreaker: question 7

A to Z are metro lines. The letters stand for stations on the line, starting at one end.

A = Asakusa (Tokyo)

B = Bakerloo (London)

C = Circle (London)

D = District (London)

E = EverLine (Seoul)

F = Fukutoshin (Tokyo)

G = Ginza (Tokyo)

H = Hyoksin (Pyongyang)

I = Imazatosuji (Osaka)

J = Jubilee (London)

K = Kakhovskaya (Moscow)

L = Leninskaya (Novosibirsk)

M = Millennium (Vancouver)

N = Northern (London)

O = Orange (Washington)

P = Piccadilly (London)

Q = Quatre (Paris)

R = Ring (Amsterdam)

S = Silver (Washington)

T = Tozai (Sapporo)

U = Urquiza (Buenos Aires)

V = Victoria (London)

W = Wenshan/Wenhu (Taipei)

X = Xinzhuang/Xinlu (Taipei)

Y = Yurakucho (Tokyo)

Z = Zamoskvoretskaya (Moscow)

279. Tiebreaker: question 8

(a) **7.** Numbers ending in EN

(b) **100.** Values of UK coins in pence, ordered by increasing coin size

(c) **4.** Word lengths of 'God Save the Queen'

(d) **60.** $p_n^2 + p_{n+1}$: 2^2+3, 3^2+5, 5^2+7, 7^2+11, 11^2+13, 13^2+17, 17^2+19, 19^2+23, ...

(e) **16.** Pope numbers from 1800, ending with Benedict XVI

(f) **8.** Smallest integers whose squares start 1, 2, 3, 4, 5, 6, 7, ...

(g) **61.** N, NNE, NE, ENE, ..., S, SSW, ..., summed using A=1, B=2, ...

(h) **18.** The completed table is:

16	?	53
53	8	7
7	92	7
32	60	68

Converting to chemical symbols using the periodic table gives:

S	?	I
I	O	N
N	U	N
Ge	Nd	Er

Ar = 18 completes this 'word rectangle' to form the words SARI and AROUND.

280. Tiebreaker: question 9

```
A M G I S G Z
H S O H I R E
S T I M U H T
A N E T P K A
D L C H S E R
H A P E I M A
E O A T L A M
Q T T A O S M
E E E U N H A
A D B M A L G
```

The following Greek and Hebrew letter names can be found in the grid: BETA, GAMMA, ZETA, ETA, THETA, IOTA, LAMBDA, SIGMA, TAU, UPSILON, GIMEL, HETH, SAMEKH, SADHE, QOPH, RESH, SHIN. When these letters are removed, the letters S, E, A, R, C, H remain in the grid; joining these letters in order traces out the letter **K**.

281. Tiebreaker: question 10

terry (or any 5-letter first name after 'penguin' in dictionary order, expressed in lower case and whose 1st letter has a Scrabble score of 1 and whose last letter corresponds to a number $\equiv$ 1 mod 4 when texted).

Scrabble score of 1st letter:

BANJUL Caius **CURT** penguin	3 points: B,C,C,P
Darren **GALAH** GONERIL győr	2 points: D,G,G,G
FRANCIS HAIFA hamlet Hawk	4 points: F,H,H,H
IAGO Norwich ORIOLE terry	1 points: I,N,O,T

Word length:

BANJUL Darren hamlet ORIOLE	6 letters
Caius **GALAH** HAIFA terry	5 letters
CURT győr Hawk **IAGO**	4 letters
FRANCIS GONERIL Norwich penguin	7 letters

Word style:

BANJUL FRANCIS GALAH IAGO	bold
Caius Darren Hawk Norwich	first letter capitalized
CURT GONERIL HAIFA ORIOLE	upper case
györ hamlet penguin terry	lower case

Texted number of last letter, modulo 4:

BANJUL GONERIL Hawk terry	L,L,K,Y ≡ 1
Caius **FRANCIS** györ ORIOLE	S,S,R,E ≡ 3
CURT **GALAH** hamlet Norwich	T,H,T,H ≡ 4
Darren HAIFA **IAGO** penguin	N,A,O,N ≡ 2

Word meaning:

BANJUL györ HAIFA Norwich	cities
Caius GONERIL hamlet **IAGO**	Shakespeare characters
CURT Darren **FRANCIS** terry	Christian names
GALAH Hawk ORIOLE penguin	birds

282. Tiebreaker: question 11

(a) **HALF PAST FOUR**. SEAHORSE/<u>H</u>IPPOCAMPINE, GOOSE/<u>A</u>NSERINE, WOLF/<u>L</u>UPINE, CAT/<u>F</u>ELINE, PIG/<u>P</u>ORCINE, EAGLE/<u>A</u>QUILINE, OWL/<u>S</u>TRIGINE, MOLE/<u>T</u>ALPINE, FINCH/<u>F</u>RINGILLINE, SHEEP/<u>O</u>VINE, BEAR/<u>U</u>RSINE, FROG/<u>R</u>ANINE

(b) **FOUR MINUTES PAST TWELVE**. The sequence is the numbers ONE, TWO, …, TWELVE enciphered via EFGHILNORSTUVWX (the letters that appear in ONE, TWO, …, TWELVE) → FOURMINTESPAWLV

(c) **1225**. Writing the three-letter groups in a column reversing alternate rows reveals the phrase HOW USA WRITES CHRISTMAS DAY USING NUMBERS.

(d) **TEN TO THREE**. 'STANDS THE CHURCH CLOCK AT
TEN TO THREE' is a line from the Rupert Brooke (3rd letters)
poem 'The Old Vicarage, Grantchester' with Grantchester
appearing in the 4th letters. FORGET, CHURCH, BARTON,
STREAM and CHERRY appear in the poem.

(e) **710**. The sequence is BRA, DPI, TTMOR, GANF, REEM, ANC,
OSTAR, BODER, EKDU, DLE, YMOORE, COSTAR with the
1st word multiplied by 1 (A=1, …, Z=26, arithmetic modulo 26),
the 2nd word multiplied by 2, and so on. This sequence can be
parsed as BRAD PITT, MORGAN FREEMAN COSTAR; BO
DEREK, DUDLEY MOORE COSTAR. Brad Pitt and Morgan
Freeman were the co-stars of SE7EN and Bo Derek and Dudley
Moore the co-stars of 10.

(f) **1605**. The sequence is REM, EMB, ERREM, EMBE, TTHE,
FIF, THOFN, OVEMB, ERBU, TWH, ICHYEA, RWASIT
with the letters in the 1st word advanced by 1 (A=1, …, Z=26,
arithmetic modulo 26), the letters in the 2nd word advanced by 2,
and so on.

(g) **1815**. Reading 1st and 2nd letters alternately reveals the lyrics
'I WAS DEFEATED YOU WON THE WAR', which appear in
ABBA's Waterloo, the eponymous battle having taken place in
1815.

(h) **1128**. The sequence is formed by alternately adding/subtracting
(A=1, …, Z=26, arithmetic modulo 26) THIRTEEN, …,
TWENTY FOUR to/from a UK speaking clock message (albeit
with a new sponsor!).

ATTHETHI RDSTROKE THETIME SPONSOR
EDBYTHEPE NGUINKRI STMASKWI ZWILLB
EELEVENTW ENTYEIGHT PRECISELYBE EPBEEPBEEP

283. Tiebreaker: question 12

(a) Boundary Undies (or any other anagram of Buried On Sunday – the last line of Solomon Grundy)

(b) Brazil Upright Haiku Mums (or any other anagram of Agh Burzum-ishi Krimpatul, the last line of the poem on the One Ring, from The Lord of the Rings)

(c) Heavyweight Tattoo Strike Fought Gout (or any other anagram of a plausible last line of: This puzzle from GCHQ, Is aimed at a brilliant few, Of the many who tried, We now must decide, [If you've got what it takes to get through])

284. Tiebreaker: question 13

The cards dealt correspond to the position of each letter in the name of a suit.

For example, S = 5♣ (5th letter of CLUBS) or 8♦ or 6♥ or 1♠ or 6♠.

The example hands are: High Card: 4♥, 3♣, 2♣, 5♠, 7♦

Pair: 5♥, 4♥, 3♣, 4♦, 2♠

Two pair: 5♥, 2♦, 4♥, 2♥, 4♠

Three of a kind: 4♣, 2♦, 4♥, 4♠, 1♠

Straight: 6♥, 5♥, 3♥, 2♦, 4♥

Flush: 8♦, 2♦, 4♦, 5♦, 6♦

Full house: 4♠, 4♥, 3♦, 4♦, 3♥

Four of a kind: 5♣, 5♥, 5♦, 6♦, 5♠

Straight flush: 6♥, 5♥, 3♥, 4♥, 2♥

In the last game the setters' hands were:

5♥, 2♥, 3♠, 4♥, 8♦ (high card)

5♥, 5♠, 3♦, 4♥, 1♠ (pair)

5♥, 5♠, 3♦, 4♥, 5♣ (three of a kind)

5♥, 2♥, 3♠, 4♥, 6♠ (straight)

5♥, 2♥, 3♥, 4♥, 6♥ (6-high straight flush)

The third setter's hand was NOMAD (indicated by the wander, desert and dry hints):

6♦, 5♦, 4♦, 3♦, 7♦ (7-high straight flush)

285. Tiebreaker: question 14

The best answer provided (by the eventual competition winner, Angus Walker) was of size 36:

A	M	U	C	K	S
D	■	G	■	■	N
V	■	H	■	Q	I
E	X	■	P	■	F
R	■	J	O	■	T
B	L	O	W	Z	Y

Answers to third colour section

Einstein's Aphorism

PATRON SAINT MISSIONS

As mentioned in the question, the answer is slightly odd!

The sudokus, once solved, can be arranged to form the surface of a cube (hinted at by the exploded diagram of a cube) so that the numbers on the two faces along an edge are the same. Establishing this allows the otherwise incomplete sudokus to be solved.
The arithmetic puzzle 'amass + priors = anoint' is a letter for number substitution of '36344 + 279574 = 315918'. Therefore '123456789'= 'npasomrti', and this substitution can be used to turn the numbers indexed in the sudoku grids into the answer phrase.

The title refers to Einstein's Aphorism 'God does not play dice' – this draws together the cubical nature of the sudokus and the religious references.

Dingbatagrams

Figure out what each picture means and then reorder to get a famous person from GCHQ's past.

(a) E, ANC (flag of the ANC), JO (the country code for Jordan for websites), LARK = JOAN CLARKE

(b) COCKS, CLIFF = CLIFF COCKS

(c) TURIN, G (in hexadecimal ASCII code), A LAN (A Local Area Network) = ALAN TURING

(d) TUT (i.e. short name for Tutankhamun), BILL, TE (in Katakana) = BILL TUTTE

(e) DENNIS, STAIR, TON, ALA (short name for Alanine) = ALASTAIR DENNISTON

Sayings

Kansas

The pictures are:

(a) Kate Moss	(b) Mick Jagger	(c) Pride
(d) A long road	(e) Home	(f) Count Dracula
(g) Miss World	(h) No turning	(i) Tango
(j) Two	(k) Archaeopteryx	(l) Cat Stevens
(m) BB King	(n) Sorry!	(o) Blessing
(p) Falls	(q) Patience	(r) Milestone
(s) Computer worm	(t) The Virtues	(u) Plaice
(v) Safe		

These represent, in various ways, words that can be paired to indicate sayings. In alphabetical order of the first of each pair, these are:

(b) A Rolling Stone gathers no Moss (a)

(c) Pride goes before a fall (p)

(d) It's a long road which has no turning (h)

(f) Count your blessings (o)

(g) A miss is as good as a mile (r)

(j) It takes two to tango (i)

(k) The early bird catches the worm (s)

(l) A cat may look at a king (m)

(q) Patience is a virtue (t)

(u) There's no place like home (e)

(v) Better safe than sorry (n)

The letters of the second word from each phrase spell out APHORISM TEN. This is 'There's no place like home', which is what Dorothy in The Wizard of Oz repeats in order to get home to the state of Kansas.

Answers to fourth colour section

People I

The pairings consist of counties and postcode diagraphs of their county towns.

(f)	Susan Hampshire	Winchester – SO	(s)	Sharon Osbourne
(g)	Leonard Cheshire	Chester – CH	(n)	Charlton Heston
(i)	Stirling Moss	Stirling – FK	(a)	Felicity Kendal
(j)	Carol Cleveland	Middlesbrough – TS	(h)	Taylor Swift
(l)	Tyrone Power	Omagh – BT	(k)	Bonnie Tyler
(o)	Victoria Derbyshire	Derby – DE	(e)	David Essex
(p)	Devon Malcolm	Exeter – EX	(u)	Evis Xheneti
(q)	Joey Essex	Chelmsford – CM	(d)	Carey Mulligan
(r)	Sarah Lancashire	Lancaster – LA	(c)	Louis Armstrong
(t)	Kiefer Sutherland	Dornoch – IV	(b)	Indira Varma
(v)	Somerset Maugham	Taunton – TA	(m)	Tori Amos

Picture sums I

Picture of a pound coin.

$$\text{Grain} \times \left[\frac{\text{chain}}{\text{rod}} + \frac{\text{yard}}{\text{foot}} \right] \times \sqrt{\frac{\text{meter}}{\text{Mike Ron}}} = \text{grain} \times (4 + 3) \times 1000 = \text{pound}$$

Picture sums II

Picture of an Aye-Aye, or an Aye-Aye eye!

Roman numerals – two answers depending on whether the 10mm×10mm×10mm cube is interpreted as a cubic centimetre (CC) or a millilitre (ML):

$$\sqrt{\sqrt{\sqrt{\left\{ MD + \frac{DC \times CC}{MCC} \right\} + IV - XL}}} = 2 = II = \text{an Aye-Aye}$$

or

$$\sqrt{\sqrt{\sqrt{\left\{ MD + \frac{DC \times ML}{MCC} \right\} + IV - XL}}} = 3 = III = \text{an Aye-Aye eye!}$$

Picture sequences

(a) r**ALPH A**rliss (British actor, of Doctor Who and The Sweeney)

catherine **ZETA**-jones

tim brook**E-TA**ylor

joh**N U**pdike

trevo**R HO**ward

guy rit**CHI**e

t**OM EGA**n (baseball player with the California Angels and Chicago White Sox)

(b) Artworks by namesakes of Teenage Mutant Ninja Turtles in alphabetical order, tinted according to mask colour: Donatello (purple), Leonardo (blue), Michelangelo (orange), so the answer would be a **Raphael tinted red**.

(c) Dub Taylor, Lin Shaye, Lis Sladen, Bon Scott, Lon Chaney, Don Cheadle, Mos Def, **?**

First names are three-letter words that join together in pairs to make capital cities: DUB–LIN, LIS–BON, LON–DON, MOS–? So the final image is of a **Cow** (or singer **'Cow-Cow' Davenport**, etc.).

Picture pairs I

(c) and (b) – **Gods and Monsters**

(d) and (g) – **Tango and Cash**

(f) and (a) – **Lilo and Stitch**

(h) and (k) – **Fast and Furious**

(j) and (e) – **Turner and Hooch**

(l) and (i) – **Q and A**

Picture connections I

Bond Girls (in order of appearance): **Honey** Ryder, **Pussy** Galore, Kissy **Suzuki, Paris** Carver, **Christmas** Jones, **Strawberry** Fields.

People II

The people share their first names with the British monarchs from James II to Elizabeth I

(a) James Nesbitt

(b) Mary Berry

(c) William Shatner

(d) Anne Hathaway

(e) George Foreman

(f) George Ezra

(g) George Eliot

(h) George Clooney

(i) Will.i.am

(j) Victoria Azarenka

(k) Edward Norton

(l) George Kennedy

(m) Edward Hopper

(n) George Harrison

(o) Elizabeth Warren

Picture sums III

Picture of actor Gordon Scott (or someone named Scott Gordon!).

Thunderbirds pilots and the numbers of their vehicles:

(Elton **John** $\pm$ **Alan** Rickman) $\div$ **Virgil** = (5 $\pm$ 3) $\div$ 2 = 4 or 1 = Gordon or Scott

Picture sums IV

Picture of a garden gate.

(Steps – Lawnmower / One little duck) × (Two little ducks / Two fat ladies)

= (39 – 14 / 2) × (22 / 88) = 8 = Garden gate

People III

All share their names with a US President (arranged in chronological order of their presidencies)

(a) Denzel **Washington**

(b) Nicola **Adams**

(c) Peter **Jackson**

(d) **Harrison** Ford

(e) Andrew **Lincoln**

(f) Hugh **Grant**

(g) **Arthur** Lowe

(h) Carol **Cleveland**

(i) Rebel **Wilson**

(j) **Truman** Capote

(k) Charles **Kennedy**

(l) Harrison **Ford**

(m) Helena Bonham-**Carter**

(n) Kate **Bush**

(o) Judd **Trump**

People IV

(a) (i) Betty Boo   (vii) Holly Hunter

    (ii) Bjorn Borg   (viii) Lyle Lovett

    (iii) Willy Brandt   (ix) James May

    (iv) Mel Brooks   (x) Liam Payne

    (v) Rose Byrne   (xi) Casey Stoney

    (vi) Donald Glover   (xii) Howard Webb

The teams are:

People whose surnames are first names of US Ryder Cup golfers (i, iii, iv, vii, x, xii):

Betty **Boo** – **Boo** Weekley

Willy **Brandt** – **Brandt** Snedeker

Mel **Brooks** – **Brooks** Koepka

Holly **Hunter** – **Hunter** Mahan

Liam **Payne** – **Payne** Stewart

Howard **Webb** – **Webb** Simpson

People whose first names are surnames of European Ryder Cup golfers ii, v, vi, viii, ix, xi):

Bjorn Borg – Thomas **Bjorn**

Rose Byrne – Justin **Rose**

Donald Glover – Luke **Donald**

Lyle Lovett – Sandy **Lyle**

James May – Mark **James**

Casey Stoney – Paul **Casey**

(b) They share their names with current UK Members of Parliament:

    (i) Chris Evans   Islwyn (Lab)

    (ii) Fiona Bruce   Congleton (Con)

    (iii) Tom Watson   West Bromwich East (Lab)

    (iv) Helen Hayes   Dulwich and West Norwood (Lab)

Picture pairs II

The pairs are bands of the form 'X and the Y', where X is a first name:

Derek (b) and the Dominoes (h)   (Derek Jacobi)

Florence (d) and the Machine (a)

Adam (e) and the Ants (j)

Martha (f) and the Muffins (l)   (Martha Washington)

Gerry (i) and the Pacemakers (g)   (Gerry Francis)

Eddie (k) and the Hot Rods (p)   (Eddie Redmayne)

Mike (m) and the Mechanics (o)

Bennie (n) and the Jets (c)   (Bennie Goodman)

'Bennie and the Jets' (sometimes also spelled 'Benny and the Jets') is the odd pair out – this is a fictional band from an Elton John song of the same name.

Wordsearch

20 cities in the International Phonetic Alphabet – 10 using the English pronunciation, and 10 using the native pronunciation.

City	Native	IPA	Grid
Algiers		ælˈdʒɪəz	G9-N9
Athens		ˈæθənz	G8-G13
Berlin		bə.ˈlɪn	B7-B1
Brussels	Bruxelles	bʁyk.sɛl	G5-N5
Cairo		ˈkaɹɹoʊ	L12-R6
Copenhagen	København	ˌkø.bənˈhavn	E9-E20
Dublin		ˈdʌblən	B16-H10
Dublin	Baile Átha Cliath	ˌbilʲɑːˈclʲiə	B22-K22
Kiev	Київ	ˈkɪjiu̯	F9-A14
Lilongwe		lɪˈlɒŋweɪ	E4-M4
Lisbon	Lisboa	liʒˈboɐ	J11-P17
London		ˈlʌn.dən	O4-H11
Minsk		mɪnsk	K10-O6
Moscow	Москва́	meskˈva	D1-D7
Paris	[French pronunciation]	paʁl	K19-N16

Rome	Roma	'ro.ma	M18-R13
Sofia		'səʊfi.ə	E16-L16
Vienna	Wien	viːn	E19-H19
Warsaw	Warszawa	var'ṣa.va	K6-C6
Zagreb		'zɑːgɹɛb	N8-N15

People V

In each case, the last three letters of the person's name is the first three letters of the next, forming a cycle.

(a) Donald Sumpter

(b) Terry Venables

(c) Leslie Ash

(d) Ashton Kutcher

(e) Herbert Hoover

(f) Vernon Kay

(g) Kaye Kendall

(h) Allison Janney

(i) Neymar

(j) Martin Luther King

(k) Ingrid Bergman

(l) Man Ray

(m) Ray Reardon

Answers to fifth colour section

Fun with flags

(a) Afghanistan, Albania, Algeria

(b) Australia, Austria, Azerbaijan

(c) Cyprus, Czech Republic, Denmark

(d) Fiji, Finland, France

(e) Indonesia, Iran, Iraq

(f) Kenya, Kiribati, Kosovo

(g) Monaco, Mongolia, Montenegro

(h) Norway, Oman, Pakistan

(i) Poland, Portugal, Qatar

(j) Yemen, Zambia, Zimbabwe

People VI

The pairings are of people sharing the same first name, except that one from each pair is better known by a pseudonym:

(a) Lee Majors (Harvey Lee Yeary) (s) Harvey Smith

(c) John Wayne (Marion Morrison) (i) Marion Bartoli

(d) Anne Rice (Howard O'Brien) (n) Howard Carter

(f) Quentin Crisp (Denis Pratt) (r) Denis Healey

(h) Annie Oakley (Phoebe Ann Mosey) (m) Phoebe Cates

(j) John le Carré (David Cornwell) (p) David Hockney

(k) Dido (Florian Armstrong) (w) Florian Schneider (-Esleben)

(l) Coco Chanel (Gabrielle Chanel) (b) Gabrielle Drake

(o) Cheryl Baker (Rita Crudgington) (g) Rita Tushingham

(q) Vin Diesel   (Mark Sinclair)   (u) Mark Zuckerberg

(t) Joni Mitchell   (Roberta Anderson)   (e) Roberta Flack

(v) Jane Seymour   (Joyce Frankenberg)   (x) Joyce Grenfell

Picture chains

(a) John Wayne, **Wayne Rooney**, Rooney Mara, **Mara Wilson**, Wilson Pickett

(b) Melanie Laurent, **Laurent Robert,** Robert Mark, Mark Ant(h)ony, **Anthony Kim, Kim Novak**, Novak Djokovic

(c) Les Paul, **Paul Nicholas, Nicholas Ridley**, Ridley Scott, **Scott Parker**, Parker Posey

(d) London Underground/Docklands Light Railway stations: Colliers Wood, **Wood Green, Green Park, Park Royal**, Royal Albert

Picture sums V

(a) **Picture of the Christmas tree emoji.**

Unicode values for emojis (hexadecimal):

(1F637 × 1F628) − (1F639 × 1F624) − 1F910 = 1F384 = Christmas tree emoji

(b) **Picture of a map showing Route 60, or Route 20.**

US Numbered Highways: the sum is:

Route 80 − (Route 220 × Route 6 / Route 66)

80 − (220 × 6 / 66) = 60 = Route 60

or, interpreting 'Route' as 'Root' gives:

$\sqrt{80} - (\sqrt{220} \times \sqrt{6} / \sqrt{66}) = \sqrt{80} - \sqrt{20} = \sqrt{20}$ = Route 20

(c) **Picture of Claire Goose.**

The Twelve Days of Christmas (using Austin's 1909 version):

Isabel **Swan** − (Billie **Piper** / (Jack **Lord** + Andy **Partridge**)) = 7 − (11 / (10+1)) = 6 = Claire Goose

People VII

(a)	Teri Hatcher	(j)	Shirley Temple
(b)	Eddie Izzard	(k)	Pablo Escobar
(c)	Natalie Imbruglia	(l)	Robert Schumann
(d)	Tony Iommi	(m)	Oscar Niemeyer
(e)	Angela Lansbury	(n)	Anita Roddick
(f)	Sharon Osbourne	(o)	Elizabeth Hurley
(g)	Frank Thornton	(p)	Andrew Neil
(h)	Harry Enfield	(q)	Dusty Springfield
(i)	Lily Allen	(r)	?

The initials of the first 17 people spell out the following message: **THE INITIALS OF THE LAST PERSON ARE H AND S**. The last person is: **Harry Styles**

Picture connections II

(a) **NBA basketball teams**: (Brooklyn) **Net**s, (Cleveland) **Cavalier**s, (Denver) **Nugget**s, (Detroit) **Piston**s, (Los Angeles) **Clipper**s, (New Orleans) **Pelican**s

(b) **Nicknames of sportspeople**: **Refrigerator** (William Perry - American football), **King of Spain** (Ashley Giles - cricket), **Hurricane** (Alex Higgins - snooker, or Rubin Carter -boxing), **Golden Balls** (David Beckham - football)

(c) **One-offs of footballers**: **Garth Brooks** (Garth Crooks), **Romney Marsh** (Rodney Marsh), **John Kerry** (John Terry), **Pole** (Pele), **Paul Merton** (Paul Merson)

(d) **NATO Phonetics with the first letter changed**: **Wolf** (Golf), **Silo** (Kilo), **Hike** (Mike), **Mango** (Tango), **Sulu** (Zulu)

People VIII

The surnames are (nearly) the same as those of the surnames of the actors that have played Doctor Who, in order.

(a) Norman Hartnell

(b) David Troughton

(c) Sean Pertwee

(d) Matt Baker

(e) Emily Davison

(f) Cheryl Baker

(g) Tony McCoy

(h) Stephen McGann

(i) William Hurt

(j) Petra Ecclestone

(k) Neil Tennant

(l) Sheridan Smith

(m) Jim Capaldi

(n) Roger Whittaker

Appendix

Periodic Table of Elements

1 IA 1A																	18 VIIIA 8A
1 H Hydrogen	2 IIA 2A											13 IIIA 3A	14 IVA 4A	15 VA 5A	16 VIA 6A	17 VIIA 7A	**2 He** Helium
3 Li Lithium	**4 Be** Beryllium											**5 B** Boron	**6 C** Carbon	**7 N** Nitrogen	**8 O** Oxygen	**9 F** Flourine	**10 Ne** Neon
11 Na Sodium	**12 Mg** Magnesium	3 IIIB 3B	4 IVB 4B	5 VB 5B	6 VIB 6B	7 VIIB 7B	8 VIII 8	9 VIII 8	10 VIII 8	11 IB 1B	12 IIB 2B	**13 Al** Aluminium	**14 Si** Silicon	**15 P** Phosphorus	**16 S** Sulfur	**17 Cl** Chlorine	**18 Ar** Argon
19 K Potassium	**20 Ca** Calcium	**21 Sc** Scandium	**22 Ti** Titanium	**23 V** Vanadium	**24 Cr** Chromium	**25 Mn** Manganese	**26 Fe** Iron	**27 Co** Cobalt	**28 Ni** Nickel	**29 Cu** Copper	**30 Zn** Zinc	**31 Ga** Gallium	**32 Ge** Germanium	**33 As** Arsenic	**34 Se** Selenium	**35 Br** Bromine	**36 Kr** Krypton
37 Rb Rubidium	**38 Sr** Strontium	**39 Y** Yttrium	**40 Zr** Zirconium	**41 Nb** Niobium	**42 Mo** Molybdenum	**43 Tc** Technetium	**44 Ru** Ruthenium	**45 Rh** Rhodium	**46 Pd** Palladium	**47 Ag** Silver	**48 Cd** Cadmium	**49 In** Indium	**50 Sn** Tin	**51 Sb** Antimony	**52 Te** Tellurium	**53 I** Iodine	**54 Xe** Xenon
55 Cs Cesium	**56 Ba** Barium	57-71	**72 Hf** Hafnium	**73 Ta** Tantalum	**74 W** Tungsten	**75 Re** Rhenium	**76 Os** Osmium	**77 Ir** Iridium	**78 Pt** Platinum	**79 Au** Gold	**80 Hg** Mercury	**81 Tl** Thallium	**82 Pb** Lead	**83 Bi** Bismuth	**84 Po** Polonium	**85 At** Astatine	**86 Rn** Radon
87 Fr Francium	**88 Ra** Radium	89-103	**104 Rf** Rutherfordium	**105 Db** Dubnium	**106 Sg** Seaborgium	**107 Bh** Bohrium	**108 Hs** Hassium	**109 Mt** Meitnerium	**110 Ds** Darmstadtium	**111 Rg** Roentgenium	**112 Cn** Copernicium	**113 Nh** Nihonium	**114 Fl** Flerovium	**115 Mc** Moscovium	**116 Lv** Livermorium	**117 Ts** Tennessine	**118 Og** Oganesson

Atomic Number | **Symbol** | Name

Lanthanide Series	**57 La** Lanthanum	**58 Ce** Cerium	**59 Pr** Praseodymium	**60 Nd** Neodymium	**61 Pm** Promethium	**62 Sm** Samarium	**63 Eu** Europium	**64 Gd** Gadolinium	**65 Tb** Terbium	**66 Dy** Dysprosium	**67 Ho** Holmium	**68 Er** Erbium	**69 Tm** Thulium	**70 Yb** Ytterbium	**71 Lu** Lutetium
Actinide Series	**89 Ac** Actinium	**90 Th** Thorium	**91 Pa** Protactinium	**92 U** Uranium	**93 Np** Neptunium	**94 Pu** Plutonium	**95 Am** Americium	**96 Cm** Curium	**97 Bk** Berkelium	**98 Cf** Californium	**99 Es** Einsteinium	**100 Fm** Fermium	**101 Md** Mendelevium	**102 No** Nobelium	**103 Lr** Lawrencium

List of US states	State abbreviations	Capital of the state	Order the state entered the union
Alabama	AL	Montgomery	22
Alaska	AK	Juneau	49
Arizona	AZ	Phoenix	48
Arkansas	AR	Little Rock	25
California	CA	Sacramento	31
Colorado	CO	Denver	38
Connecticut	CT	Hartford	5
Delaware	DE	Dover	1
Florida	FL	Tallahassee	27
Georgia	GA	Atlanta	4
Hawaii	HI	Honolulu	50
Idaho	ID	Boise	43
Illinois	IL	Springfield	21
Indiana	IN	Indianapolis	19
Iowa	IA	Des Moines	29
Kansas	KS	Topeka	34
Kentucky	KY	Frankfort	15
Louisiana	LA	Baton Rouge	18
Maine	ME	Augusta	23
Maryland	MD	Annapolis	7
Massachusetts	MA	Boston	6
Michigan	MI	Lansing	26
Minnesota	MN	Saint Paul	32
Mississippi	MS	Jackson	20
Missouri	MO	Jefferson City	24
Montana	MT	Helena	41
Nebraska	NE	Lincoln	37
Nevada	NV	Carson City	36
New Hampshire	NH	Concord	9
New Jersey	NJ	Trenton	3
New Mexico	NM	Santa Fe	47
New York	NY	Albany	11
North Carolina	NC	Raleigh	12
North Dakota	ND	Bismarck	39

Ohio	OH	Columbus	17
Oklahoma	OK	Oklahoma City	46
Oregon	OR	Salem	33
Pennsylvania	PA	Harrisburg	2
Rhode Island	RI	Providence	13
South Carolina	SC	Columbia	8
South Dakota	SD	Pierre	40
Tennessee	TN	Nashville	16
Texas	TX	Austin	28
Utah	UT	Salt Lake City	45
Vermont	VT	Montpelier	14
Virginia	VA	Richmond	10
Washington	WA	Olympia	42
West Virginia	WV	Charleston	35
Wisconsin	WI	Madison	30
Wyoming	WY	Cheyenne	44

List of US presidents:

1.	George Washington	16.	Abraham Lincoln	31.	Herbert Hoover
2.	John Adams	17.	Andrew Johnson	32.	Franklin D. Roosevelt
3.	Thomas Jefferson	18.	Ulysses S. Grant	33.	Harry S. Truman
4.	James Madison	19.	Rutherford B. Hayes	34.	Dwight D. Eisenhower
5.	James Monroe	20.	James A. Garfield	35.	John F. Kennedy
6.	John Quincy Adams	21.	Chester A. Arthur	36.	Lyndon B. Johnson
7.	Andrew Jackson	22.	Grover Cleveland	37.	Richard Nixon
8.	Martin Van Buren	23.	Benjamin Harrison	38.	Gerald Ford
9.	William Henry Harrison	24.	Grover Cleveland	39.	Jimmy Carter
10.	John Tyler	25.	William McKinley	40.	Ronald Reagan
11.	James K. Polk	26.	Theodore Roosevelt	41.	George Bush
12.	Zachary Taylor	27.	William Howard Taft	42.	Bill Clinton
13.	Millard Fillmore	28.	Woodrow Wilson	43	George W. Bush
14.	Franklin Pierce	29.	Warren G. Harding	44	Barack Obama
15.	James Buchanan	30.	Calvin Coolidge	45	Donald Trump

NATO phonetic alphabet:

A	–	Alpha/Alfa	N –	November
B	–	Bravo	O –	Oscar
C	–	Charlie	P –	Papa
D	–	Delta	Q –	Quebec
E	–	Echo	R –	Romeo
F	–	Foxtrot	S –	Sierra
G	–	Golf	T –	Tango
H	–	Hotel	U –	Uniform
I	–	India	V –	Victor
J	–	Juliet/Juliett	W –	Whiskey
K	–	Kilo	X –	X-Ray
L	–	Lima	Y –	Yankee
M	–	Mike	Z –	Zulu

Greek alphabet:

Alpha	Nu
Beta	Xi
Gamma	Omicron
Delta	Pi
Epsilon	Rho
Zeta	Sigma
Eta	Tau
Theta	Upsilon
Iota	Phi
Kappa	Chi
Lambda	Psi
Mu	Omega

Morse code, letters and numbers:

A	• —	N	— •	0	— — — — —
B	— • • •	O	— — —	1	• — — — —
C	— • — •	P	• — — •	2	• • — — —
D	— • •	Q	— — • —	3	• • • — —
E	•	R	• — •	4	• • • • —
F	• • — •	S	• • •	5	• • • • •
G	— — •	T	—	6	— • • • •
H	• • • •	U	• • —	7	— — • • •
I	• •	V	• • • —	8	— — — • •
J	• — — —	W	• — —	9	— — — — •
K	— • —	X	— • • —		
L	• — • •	Y	— • — —		
M	— —	Z	— — • •		

Scrabble table of values and frequencies

2 blank tiles   (scoring 0 points)

1 point: E×12, A×9, I×9, O×8, N×6, R×6, T×6, L×4, S×4, U×4

2 points: D×4, G×3

3 points: B×2, C×2, M×2, P×2

4 points: F×2, H×2, V×2, W×2, Y×2

5 points: K×1

8 points: J×1, X×1

10 points: Q×1, Z×1

Semaphore alphabet:

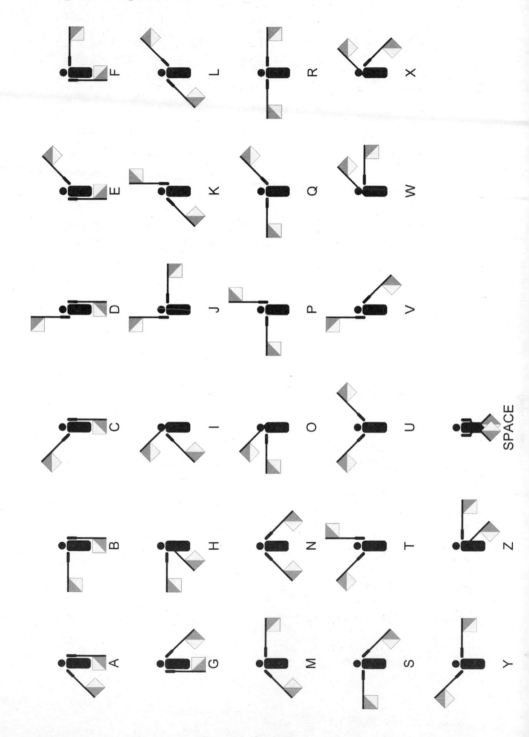

Braille alphabet:

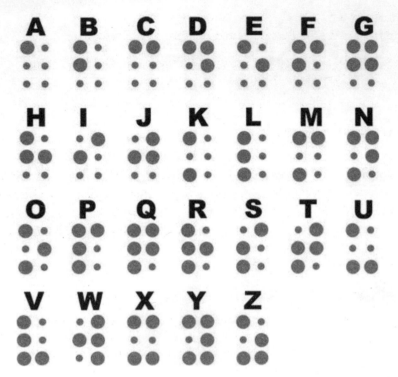

Picture Credits

Picture pairs II: (a) Kontrola/Wikimedia/CC BY-SA 3.0; (b) Airman 1st Class Derek Seifert/U.S. Air Force/Wikimedia/Public Domain; (c) Florian Decker, Messdiener Winterbach/Wikimedia/CC BY-SA 3.0; (d) The Yorck Project (2002)/Wikimedia/Public Domain; (e) Library of Congress Prints and Photographs Division/LC-USZ62-113386; (f) Munshi Ahmed/Bloomberg via Getty Images; (g) Eitan f/Wikimedia/CC BY-SA 3.0; (h) Adam Fradgley - AMA/West Bromwich Albion via Getty Images; (i) United States Department of Agriculture/ k5388-1/Wikimedia; (j) Dick Thomas Johnson/Wikimedia/ CC BY-SA 2.5; (k) veganbaking.net/Wikimedia/CC BY-SA 2.0; (l) Felipe Micaroni Lalli/Wikimedia/CC BY-SA 2.5; (m) Hans Bernhard (Schnobby)/ Wikimedia/CC BY-SA 3.0; (n) Arild Vågen/Wikimedia/CC BY-SA 4.0; (o) © Shutterstock; (p) Arild Vågen/Wikimedia/CC BY-SA 4.0

People V: (a) Tim P. Whitby / Stringer; (b) Chris Jackson/Getty Images; (c) Andy Butterton/PA Archive/PA Images; (d) David Shankbone/Wikimedia/ CC BY 2.0; (e) Library of Congress Prints and Photographs Division/ Underwood & Underwood/ LC-USZ62-24155; (f) © Shutterstock; (g) AF archive / Alamy Stock Photo; (h) Mingle Media TV/Wikimedia/CC BY-SA 2.0; (i) Granada/Wikimedia/CC BY-SA 4.0; (j) Stamptastic / Shutterstock. com; (k) John Kobal Foundation / Contributor / Getty images; (l) Granger Historical Picture Archive / Alamy Stock Photo; (m) DAVID MUSCROFT / Alamy Stock Photo

Fifth Colour Section

Fun with flags: All images from Wikimedia

People VI: (a) © Shutterstock; (b) Popperfoto/Getty Images; (c) AF archive / Alamy Stock Photo; (d) Johnny Louis/Getty Images; (e) © Shutterstock; (f) © Shutterstock; (g) Petr Novák, Wikipedia/CC BY-SA 2.5; (h) © Shutterstock; (i) Clive Brunskill/Getty Images; (j) © Shutterstock; (k) Dave Hogan/Getty Images; (l) Pictorial Press Ltd / Alamy Stock Photo; (m) © Shutterstock; (n) Library of Congress Prints and Photographs/ National Photo Company Collection/LC-F8- 30481; (o) © Shutterstock; (p) Hulton-Deutsch Collection/CORBIS/Corbis via Getty Images; (q) © Shutterstock; (r) © Shutterstock; (s) Wikimedia; (t) Wikimedia; (u) Mika-photography/ Wikimedia/CC BY-SA 3.0; (v) © Shutterstock; (w) Daniele Dalledonne/ Wikimedia/CC BY-SA 2.0; (x) Allan Warren/Wikimedia/CC BY-SA 3.0

Credits

The puzzles in this book were created by the staff of GCHQ in their spare time.

Puzzle editor: Colin

Deputy puzzle editor: Chris

Assistant puzzle editors: Andrew, Daniel, Mike, Richard

Principal puzzle setters: Andrew, Celia, Chris, Colin, Daniel, John, Julian, Katy, Mike, Nigel, Richard, Tom, Tom 2, Will

Additional puzzles set by: Andrew 2, Andy, Andy 2, Carol, Flo, Hannah, Ian, Ian 2, James, James 2, Jamie, Michael, Nick, Paul, Peter, Rob, Robert, Robin, Sam, Samantha, Tom 3, Will 2

Thanks also to: Joanna, Kirsti, Linda

Historical interludes: Daniel, Maddy, Stuart, Tony

Particular thanks go to Fi, and to Amy, Amy 2, Andrew 3, Faye, Fi 2, Matt and Wayne

Thanks also to the PFD agency, Dan Prescott for text design, and to the team at Penguin: Daniel Bunyard, Clare Parker, Emma Henderson, Amy McWalters, Annie Underwood, Alice Chandler, Ellie Hughes

Strenuous efforts have been made by both Penguin
and GCHQ to ensure that this book is free from errors.
However, it contains a lot of intricate detail and it would be
arrogant of us to assert that we have completely succeeded.
If you think you have found an error, please email:

GCHQ2@penguinrandomhouse.co.uk